Orbitals

With Applications in Atomic Spectra

ICP Essential Textbooks in Chemistry

Vol. 1: Orbitals: With Applications in Atomic Spectra
by Charles Stuart McCaw (Winchester College, UK)

ICP Essential Textbooks in Chemistry

Orbitals

With Applications in Atomic Spectra

Charles S. McCaw
Winchester College, UK

Imperial College Press
ICP

Published by

Imperial College Press
57 Shelton Street
Covent Garden
London WC2H 9HE

Distributed by

World Scientific Publishing Co. Pte. Ltd.
5 Toh Tuck Link, Singapore 596224
USA office: 27 Warren Street, Suite 401-402, Hackensack, NJ 07601
UK office: 57 Shelton Street, Covent Garden, London WC2H 9HE

Library of Congress Cataloging-in-Publication Data
McCaw, Charles Stuart.
Orbitals : with applications in atomic spectra / by Charles Stuart McCaw, Winchester College, UK.
pages cm
Includes bibliographical references and index.
ISBN 978-1-78326-413-1 (hardcover : alk. paper)
1. Molecular orbitals. 2. Atomic spectra. 3. Chemistry, Physical and theoretical. I. Title.
QD461.M396 2015
541'.28--dc23
2015000745

British Library Cataloguing-in-Publication Data
A catalogue record for this book is available from the British Library.

Typeset by Stallion Press
Email: enquiries@stallionpress.com

Printed in Singapore

This book is dedicated to the memory of Bob Denning (1938–2013), my D. Phil. supervisor and a constant source of inspiration.

Contents

Acknowledgements

I am grateful to Laurent Chaminade of Imperial College Press and Richard Compton of Oxford University for taking an interest in this project, to Catharina Weijman of Imperial College Press for her patience with me, and to Winchester College for granting me a reduction in duties to complete the book.

The author gratefully acknowledges the detailed comments on the manuscript by Eugen Schwarz (University of Siegen and Tsinghua University, Beijing), and useful discussions with Eric Scerri (University of California, Los Angeles), Bob Denning (University of Oxford), John Cullerne, David Follows and Geoffrey Eyre (all Winchester College), Jeremy Burrows (The Perse School), Tim Greene (Clifton College), Oliver Choroba (Charterhouse School) and Madeleine Copin (North London Collegiate School). The author is grateful to Roberto Faria (Universidade Federal do Rio de Janeiro) for providing Rich and Suter diagrams specially for this book. Thanks too are due to Julian Perry (Winchester College) for checking the manuscript, though any remaining errors are my responsibility. Special thanks to my wife Govya for her encouragement and support of my writing.

The author used Gnuplot freeware to construct the various kinds of plot and is grateful for the help provided by a couple of sources [1,2]. The book was written by the author using LaTeX freeware, with the assistance of two useful books [3,4].

Acknowledgements

Preface

This book aims to present a modern and mathematically rigorous approach to orbitals in chemistry and apply the concepts to atomic spectra. In the process it should provide a basic introduction to quantum chemistry. It is intended for the book to have wide appeal: orbitals command broad interest across the range of students, teachers and researchers of chemistry in different contexts and at different levels of sophistication.

Electron configurations of transition metals are treated in more detail than in many other books, since this topic raises many interesting and subtle questions about orbitals, starting at the early stages of a UK sixth-form course in chemistry. Rich and Suter diagrams are presented as a useful pedagogical tool, as they provide a more intuitive simple explanation than what is typically seen in textbooks. The limitations of the aufbau principle and of using free atoms to describe electron configurations are discussed and the configuration-average approach of Schwarz is presented. These approaches are little-known and are described here to give a fresh perspective on transition metals; they should be of interest to educators.

There is a discussion of recent research into the visualisation of orbitals, which should be of interest to researchers with an interest in electronic structure. The chapter on orbitals in molecules considers some less common geometries of overlap, and gives a detailed view of π bonding in benzene. The chapter on orbitals in the hydrogen atom considers the f orbitals in both cubic and spherical symmetry. Relativistic effects seen in the s electrons of heavy atoms are explained as far as possible while remaining accessible to most chemists, i.e. without solving the Dirac equation. Electron affinity is considered in the light of data and research that is more recent than some common data books and the approach taken in sixth-form courses.

On first encountering the Schrödinger equation and the mathematical functions describing orbitals of different kinds, many chemistry students are soon confused. Some authors, wanting to protect their readers from mathematics, present only the most important results, but not knowing their origins can leave the reader unsatisfied. Other authors will present the subject pictorially and communicate qualitative aspects this way. This book aims to present the origins and some important applications of orbitals in as complete a way as possible while explaining everything step-by-step from first principles. The details are either presented in full or left to exercises, which have worked solutions at the back of the book.

Following the recent claim for the discovery of an element with an atomic number of about 122 [5] which, following the aufbau principle and recent relativistic calculations [6], might have a $g_{7/2}$ electron in its ground state configuration, g orbitals are dealt with in this book (albeit in the somewhat idealised sense in a hydrogen atom). They are not, however, presented entirely alongside the better-known atomic orbitals, as they are only of theoretical interest. In this light there is some consideration of the limits of the periodic table and the relativistic factors involved.

Where possible, mathematical group theory is avoided in this book. In keeping with the aims of satisfying the reader that each step is explained in full, examples are chosen to illustrate calculations that do not require group theory. While the author acknowledges the power and elegance of group theory, for many chemists it is a means of arriving at answers to problems rather than a vehicle for gaining deeper understanding.

With over 100 worked solutions to exercises in the appendix, this book should also serve as a tool for learning how to solve problems involving orbitals and atomic spectra, as well as being a textbook.

Most of the material in this book has been taught at various times to able sixth-form students at Winchester College with good training in mathematics; I am grateful to them for their ideas and questions, which have helped in the writing of this book. I believe therefore that it should be accessible to undergraduate chemistry students.

Chapter 1

Fundamentals

When school students first meet electrons they are considered to be orbiting a nucleus like planets orbit the sun, i.e. with classical mechanics. Here the quantum mechanical (or wave mechanical) description is presented. It is worth bearing in mind that — even with the mathematical refinement of this treatment — this is not the final word: no account is taken of relativistic effects or quantum effects in the vacuum, though for valence electrons in light atoms these effects are very small.

1.1 Wave–Particle Duality

At the root of it all is wave–particle duality,wave–particle duality which is an experimental fact: electrons can impart momentum in collisions like particles do and they can be diffracted and produce interference patterns like waves do. Whether they behave as waves or particles depends on the experiment.[1] For our purposes we are taking electrons in atoms to behave as standing waves. By analogy, a plucked guitar string holds a standing wave. We can imagine the string to be a one-dimensional object. The two-dimensional analogy might be the flat surface of a drum being struck. The three-dimensional analogy might

[1]The modern quantum-physics approach considers the electron as a matter field represented by a wavefunction that always gives a distribution for position or velocity, even when the electron seems to be behaving as a particle. So the most localised position of an electron is a spot rather than a point.

be a rubber ball vibrating around its own centre. Electron wavefunctions have three spatial dimensions.

Paul Dirac argued in his classic monograph on quantum mechanics (see the first section in [7]) that wave–particle duality could be justified on philosophical grounds. Considering the experimental fact that matter is not continuous but composed of fundamental particles, it follows that there must be some degree of indeterminacy at the level of these particles, since any observation made of a fundamental particle must involve disturbing it (with a photon, for example). In describing an electron as a wave we have introduced the required indeterminacy in the simultaneous measurement of its position and momentum.

1.2 Wavefunctions

The Greek letter psi (ψ) is normally used to represent the wavefunction of an electron. We will see later that the wavefunction is inherently complex, i.e. containing the imaginary i. We will begin by considering the real functions that describe classical waves.

A general formula for the amplitude at a given point on the x-axis of a classical wave travelling from right to left is

$$\psi = A \sin(\omega t + \phi), \tag{1.1}$$

where A is the amplitude of the wave at the given point, ω is the angular frequency, t is time elapsed and ϕ is the phase of the wave in radians at $t = 0$. The angular frequency is defined as

$$\omega = 2\pi f, \tag{1.2}$$

where f is the frequency, e.g. in hertz (s^{-1}). ωt is therefore 2π times the number of wavelengths that passed the point in time t. It is therefore the phase of the wave in radians at time t when $\phi = 0$.

A general formula for a stationary wave on the x-axis is

$$\psi = A \sin(kx + \phi), \tag{1.3}$$

where ϕ is the phase of the wave in radians at $x = 0$ and k is the wavenumber. Wavenumber is defined as

$$k = \frac{2\pi}{\lambda}, \tag{1.4}$$

where λ is the wavelength. This means that kx is 2π times the number of multiples of the wavelength along the x-axis. k is therefore the spatial equivalent of the angular frequency, meaning that kx is a spatial equivalent of angular phase, and is also measured in radians. We can combine equations (1.1) and (1.3) to describe a classical wave travelling in one dimension from right to left on the x-axis in terms of both space and time coordinates,

$$\psi = A\sin(kx + \omega t + \phi), \tag{1.5}$$

where ϕ is the phase of the wave in radians at $x = 0$ and $t = 0$.

We could also have written equation (1.5) with a cosine function if we had adjusted the value of ϕ by $\pi/2$. Since there are multiple phases in the argument of the trigonometric function it becomes mathematically convenient to express the wavefunction as a complex exponential function, which is related to trigonometric functions through de Moivre's equation,

$$\mathrm{e}^{\mathrm{i}\phi} = \cos\phi + \mathrm{i}\sin\phi, \tag{1.6}$$

where $\mathrm{i} = \sqrt{-1}$. The complex exponential form is useful because the expression can be written as the product of complex exponentials of the component phases:

$$A\mathrm{e}^{\mathrm{i}(\phi_1+\phi_2+\phi_3)} = A\mathrm{e}^{\mathrm{i}\phi_1}\mathrm{e}^{\mathrm{i}\phi_2}\mathrm{e}^{\mathrm{i}\phi_3}. \tag{1.7}$$

We will see later that the wavefunctions for electrons are necessarily complex. However, the probability density functions used to visualise where an electron may be found are necessarily real. There are mathematical processes (described later) that turn the complex wavefunctions into real probability densities. For example,

$$A\mathrm{e}^{\mathrm{i}\phi} + A\mathrm{e}^{-\mathrm{i}\phi} = 2A\cos\phi, \tag{1.8}$$

since $\cos\phi = \cos(-\phi)$ but $\sin\phi = -\sin(-\phi)$. The properties of complex exponentials and their use in describing oscillating functions are explained in [8].

1.3 Schrödinger and his Equation

When Schrödinger first constructed his wavefunction in 1926 to describe fundamental particles he wrote it in terms of dynamical variables that might be measured: position, momentum and energy. Being a non-relativistic equation, mass is treated as a constant, and time as a parameter rather than a variable. We shall be concerned with stationary states principally in this book, and so will be considering Schrödinger's time-independent wave equation. He combined the dynamical variables with the Planck constant, h, since this was a measure of the smallest packets or quanta of energy that could be measured, as illustrated by the well-known equation relating the energy, E, of a photon of light to its frequency, f,

$$E = hf. \tag{1.9}$$

Equation (1.9) shows that Planck's constant has units of J s. These are the same units as momentum $\times$ distance and also energy $\times$ time. A generic wavefunction, comparable to equation (1.5) but expressed in the form of a complex exponential is

$$\psi = A\exp\left\{2\pi\mathrm{i}\left(\frac{px - Et}{h} + \phi\right)\right\}. \tag{1.10}$$

There is a negative sign in front of the Et term as the wave is defined as travelling in the positive direction. (In equations (1.1) and (1.5) the wave is travelling in the negative direction.)

Schrödinger's famous equation is an eigenvalue equation. Such equations take the general form

$$\hat{O}\psi = \lambda\psi, \tag{1.11}$$

where $\hat{O}$ is an operator (as indicated by the hat symbol). An operator is something that acts on a function to produce another function. In an eigenvalue equation, when the operator acts on the function, called an eigenfunction, it returns the same function multiplied by a constant, λ, known as the eigenvalue. Schrödinger constructed operators that generated eigenvalues that correspond to physical observables.

The operator for momentum along the x-axis, $\hat{p}_x$, is $\frac{h}{2\pi\mathrm{i}} \times \frac{\partial}{\partial x}$. The partial differential operator $\frac{\partial}{\partial x}$ denotes differentiating with respect to x while keeping other variables constant. Dividing Planck's constant by distance is dimensionally consistent with momentum. Dividing Planck's constant by 2π takes into account the wave nature of the wavefunction, which is demonstrated in equation (1.12) below. Applying this operator to the generic wavefunction, $\hat{p}_x\psi$, returns the momentum, p, as the eigenvalue:

$$\begin{aligned}\hat{p}_x\psi &= \frac{h}{2\pi\mathrm{i}} \times \frac{\partial\psi}{\partial x} \\ &= \frac{h}{2\pi\mathrm{i}} \times \frac{\partial}{\partial x}\left\{A\exp\left(2\pi\mathrm{i}\left(\frac{px - Et}{h} + \phi\right)\right)\right\} \\ &= pA\exp\left(2\pi\mathrm{i}\left(\frac{px - Et}{h} + \phi\right)\right) = p\psi.\end{aligned} \tag{1.12}$$

This is known as the position representation, as it is the position variable being operated on. The operator for position in this representation is trivially $\times x$.

These operators for momentum and position are used to construct the Schrödinger equation in the following sections. This field is known as wave mechanics.

1.4 Properties of Eigenvalue Equations

These equations have certain properties of which chemists can take advantage:

(1) An eigenfunction may be multiplied by any constant and it remains an eigenfunction that produces the same eigenvalue from the operator.
(2) Provided that several different eigenfunctions have the same eigenvalue, then a linear combination, i.e. a sum, of eigenfunctions is also an eigenfunction of the operator. In this sense eigenfunctions add up rather like vectors and so wavefunctions are often called eigenvectors. Using the arbitrary constants from the previous point, our eigenvectors can effectively all be normalised to the same length.
(3) A certain class of operator is particularly useful for quantum chemistry, namely Hermitian operators. When operating on a linear combination of n eigenvectors an $n \times n$ matrix is required. Such a matrix is Hermitian if it is self-adjoint. A self-adjoint matrix is equal to its conjugate transpose, which is formed by reflecting all the matrix elements in the leading diagonal and taking the complex conjugate of each element that has been reflected. The reason Hermitian operators are so useful is that their eigenvalues are always real, a requirement for physical observables.
(4) Another property of Hermitian operators is that eigenvectors belonging to different eigenvalues are mutually orthogonal. Two vectors are orthogonal when their scalar, i.e. dot, product is zero. The analogous procedure with wavefunctions is to integrate the product of two wavefunctions over all space. This fact provides much simplification to quantum chemical calculations.

The properties of Hermitian operators are derived and discussed in detail in [9] and [10].

Exercise 1

Show that the (2×2) matrix below is Hermitian by pre- and post-multiplying it by (1×2) and (2×1) complex vectors. a, b, c, d, x, and y are real:

$$\begin{bmatrix} a - ib & c - id \end{bmatrix} \begin{bmatrix} m & x + iy \\ x - iy & n \end{bmatrix} \begin{bmatrix} a + ib \\ c + id \end{bmatrix}. \tag{1.13}$$

Given that the eigenvectors of a Hermitian operator are orthogonal, it is instructive to construct a series of functions, Φ, in a variable, x say, of ever increasing power of x that are all mutually orthogonal. This can be achieved a procedure known as Gram–Schmidt orthogonalisation. We could begin with $x^0 = 1$ as the first member of our orthogonal series and normalise it between -1 and 1 by setting the integral with respect to x of its square between -1 and 1 to unity:

$$\begin{aligned} N_0^2 \int_{-1}^{1} 1 \, \mathrm{d}x &= 1 \\ N_0^2 \times 2 &= 1 \\ N_0 &= \frac{1}{\sqrt{2}}. \end{aligned} \tag{1.14}$$

The first orthonormalised member of the series is therefore $\Phi_0 = \frac{1}{\sqrt{2}}$. For the next member of the series, we trial $\Phi_1 = x + a_{1,0}\Phi_0$, where $a_{1,0}$ is a constant to be determined. First we impose orthogonality:

$$\int_{-1}^{1} \Phi_1 \Phi_0 \, \mathrm{d}x = 0 = \int_{-1}^{1} \frac{x}{\sqrt{2}} \, \mathrm{d}x + a_{1,0} \int_{-1}^{1} \Phi_0 \Phi_0 \, \mathrm{d}x$$

$$a_{1,0} = -\frac{1}{\sqrt{2}} \int_{-1}^{1} x \, \mathrm{d}x = 0. \tag{1.15}$$

This leaves $\Phi_1 = N_1 x$, where N_1 is a normalisation constant to be determined, as follows:

$$N_1^2 \int_{-1}^{1} x^2 \, \mathrm{d}x = 1$$

$$N_1^2 \left[\frac{x^3}{3} \right]_{-1}^{1} = 1$$

$$N_1^2 \times \frac{2}{3} = 1$$

$$N_1 = \sqrt{\frac{3}{2}} = \frac{\sqrt{6}}{2}. \tag{1.16}$$

The second orthonormalised member of this series is therefore

$$\Phi_1 = \frac{\sqrt{6}}{2} x.$$

The process becomes increasingly laborious since, with each higher power of x, there is an addition constant, a, to determine. This is achieved by requiring each function to be orthogonal to all the lower functions. If this process is repeated it generates the Legendre polynomials. By substituting $\cos\theta$ for x, these become the associated Legendre functions, P_l^0, used in the next chapter to describe orbitals as a function of the colatitude, θ.

1.5 The Meaning of the Wavefunction

Quantum mechanics — dynamics on the smallest scales — raises many perplexing questions. One is on the physical meaning of the wavefunction itself. As Dirac wrote (in the first part of [7]), it is "important to remember that science is concerned only with observable things". The wavefunction itself is not a quantum mechanical observable (though we see how they may be constructed from observations in the next section), so one might question the validity of employing them. However, predictions of observable quantities made using wavefunctions have been confirmed experimentally countless times. Despite the philosophical problems with quantum mechanics it is arguably science's most tested and successful theory. A recent paper has claimed that wavefunctions are in fact a reality [11]. It argues that any model in which a quantum state represents mere information about an underlying physical state of the system, and in which systems that are prepared independently have independent physical states, must make predictions that contradict those of quantum theory.

Complex wavefunctions are transformed into real probability densities by multiplication by their complex conjugate, denoted ψ^*, which is the wavefunction with the sign of the imaginary component reversed. This can be shown by writing a general complex function as $(a + ib)$, where a and b

are real:

$$(a + ib)(a - ib) = a^2 - i^2b^2 = a^2 + b^2, \quad (1.17)$$

which is real. A wavefunction is a mathematical function, typically of three spatial coordinates for standing electron waves in atoms. The arbitrary constant that belongs to each wavefunction is chosen such that the function $\psi^*\psi$ integrated over all space gives 1. Thus, when the wavefunction $\psi^*\psi$ is integrated over a range of spatial coordinates, the resulting value is the probability of the electron being found in that volume. The wave nature of the electron implies that the position of the electron at any given instant is uncertain, and so only probabilities of position can be obtained with the wavefunction.

1.6 What are Orbitals and can they be Visualised?

Orbital is evidently a derivative of *orbit*, the suffix no doubt to distinguish it from the classical orbits of planets, etc. At an introductory level, orbitals are often understood to be a region of space in which an electron is likely to be found or, to put it more precisely, the three-dimensional probability density functions, $\psi^*\psi$, that describe the location of an electron in that state. As we shall see in Section 2.6, the radial functions decay asymptotically to zero as r increases. If we want to describe an orbital with a three-dimensional shape we have to decide what proportion of the function to show. Typically 90% of the probability density function is used, i.e. r is limited to the value such that 90% of the volume of the function (which, since it is defined as a probability, has a total value of 1) is shown. Such visualisations are not strictly probability densities, wavefunctions or orbitals, but the boundaries of the 3D function. These boundaries are known as isosurfaces since the wavefunction has a common value at all points on this surface.

It is now general usage, however, to consider orbitals to be one-electron wavefunctions. The term was coined in this way by Robert Mulliken in 1932 [12]. When the orbital wavefunction includes a spin function, it is often referred to as a spin-orbital. *Wavefunction* is used more generally for eigenfunctions of the Hamiltonian operator that include multi-electron states. It follows from this definition of orbitals that they do not have an independent existence in multi-electron atoms (see below).

Models and pictures of orbitals commonly involve lobes of different colour. These represent the phase or sign of the wavefunction in the different lobes. In isolated atoms the sign is lost when the probabilty density function of the wavefunction is constructed. It is significant, however, when orbitals on different atoms combine to make chemical bonds: there will be cancellation where orbitals lobes of opposite sign overlap, for example. This point will be considered further in Chapter 5. In this light, orbitals are usually considered by chemists to be the wavefunction before it is multiplied by its complex conjugate, which has been discussed by Mulder [13].

Observable properties such as the energy or momentum of an electron are obtained from projecting the wavefunction onto eigenstates. However, each projection only reveals a portion of the wavefunction; typically phase information is lost. The orbitals themselves are not, strictly, therefore quantum mechanical observables. The full quantum state is only obtained by statistically averaging over many measurements [14]. The visualisation of hydrogen orbitals was achieved in 2013 by sampling many atoms using photoionisation microscopy of the atoms. This was done in a DC electric field as hydrogen atoms — unlike molecules — have no axis of quantisation. An electrostatic lens magnified the outgoing electron wave without disrupting its wavefunction. The resulting interference pattern showed the nodal structure of the excited electron states on the atom that had been created by a tunable laser pulse [15].

A paper that made the front page of *Nature* in 1999 claimed to have observed a d_{z^2} orbital on copper in Cu_2O; the authors used a combination of convergent-beam electron diffraction and X–ray diffraction [16]. Scerri has raised the objection that the d_{z^2} orbital is only an eigenfunction in one-electron atoms (and only then in the non-relativistic approximation); in many-electron atoms these orbitals are only part of determinantal wavefunctions in the orbital approximation where no account is taken of electron correlation [17]. He concludes that hydrogenic orbitals may not in principle be observed in multi-electron atoms. This may be formalised in terms of the commutation relations. In quantum mechanics, orbitals are characterised by their angular momenta. For a one-electron orbital to be observable in an atom in a given energy level the operator for the angular momentum of the electron in that orbital, $\hat{l}_z$, and the operator for energy, the Hamiltonian, $\hat{H}$, must commute, i.e.

$$\hat{l}_z\hat{H} - \hat{H}\hat{l}_z = [\hat{l}_z,\ \hat{H}] = 0. \tag{1.18}$$

The above expression holds for orbitals in one-electron atoms but not for those in multi-electron atoms. A detailed commentary on the Cu_2O debate and explanation of the technicalities have been provided by Wang and Schwarz [18,19].

The reason that the simultaneous position and momentum of an electron may not in principle be known exactly is because the commutator $[\hat{x},\ \hat{p_x}] \neq 0$. This led Heisenberg to his famous uncertainty principle. In time-dependent processes, the energy and time coordinate of the system may also in principle not be precisely known simultaneously.

Exercise 2

Show that the position and momentum of an electron may not in principle be known simultaneously by working out the commutator $[\hat{x},\ \hat{p_x}]$.

Exercise 3

Let two operators, $\hat{A}$ and $\hat{B}$, simultaneously operate on an eigenfunction, ψ, generating eigenvalues a and b, respectively. Show that the operators, $\hat{A}$ and $\hat{B}$ must commute.

Since the claim to have observed a d_{z^2} orbital on copper, it was reported in 2004 that phase as well as amplitude information can be imaged on the highest occupied molecular orbital of a diatomic molecule [20]. This was obtained from tomographic reconstruction using intense femtosecond laser pulses, together with theoretical modelling. Submolecular imaging of surfaces has also been obtained using scanning probe techniques [21]. A development in 2014 is the more general imaging of both phase and amplitude information in adsorbed molecules using angle-resolved photoemission spectroscopy [22]. In this case the phase information is recovered via an iterative procedure that involves a reverse Fourier transform together with the assumption that the wavefunction is confined to the van der Waals size of the molecule. Significantly, the iterative process leads to the same wavefunction even when a random choice is made for the initial phase. Furthermore the wavefunctions on a molecule generated in this technique are orthogonal with one another and show remarkable agreement with the theoretical wavefunctions.

It is not clear that a one-electron wavefunction should result from the photoemission process going from an N-electron state to an $N-1$-state: strictly it should be a Dyson orbital, which represents the overlap between the initial and final wavefunctions, before and after the ionisation [23].

However, when computed for isolated molecules, these Dyson orbitals have been shown closely to resemble one-electron initial-state orbitals [24].

1.7 Dirac Notation

Multiplication of a wavefunction by its complex conjugate and integration over all space is such a common operation in quantum chemistry that Dirac developed a shorthand notation used by all in the field. The integral of $\psi^*\psi$ over all space is abbreviated as

$$\langle\psi|\psi\rangle = \int_{-\infty}^{\infty}\int_{-\infty}^{\infty}\int_{-\infty}^{\infty} \psi^*(x,y,z)\psi(x,y,z)\,\mathrm{d}x\,\mathrm{d}y\,\mathrm{d}z, \tag{1.19}$$

where $\langle\psi|$ is known as the bra vector, and $|\psi\rangle$ as the ket vector. The bra vector is a complex conjugate. If the bra and ket are both real functions then $\langle A|B\rangle = \langle B|A\rangle$, otherwise

$$\langle A|B\rangle = \langle B|A\rangle^*. \tag{1.20}$$

Exercise 4

Show that $\langle A|B\rangle = \langle B|A\rangle^*$. Use $|A\rangle = a + ib$ and $|B\rangle = x + iy$.

Expressions involving operators can also be expressed with Dirac bracket notation. For example, equation (1.11) is written as

$$O\mid\psi\rangle = \lambda|\psi\rangle. \tag{1.21}$$

It is only when the bra vector is included that an integration over all space is implied. This is required for solving eigenvalue equations, as follows. We insert to the left-hand side of each half of the above equation the bra vector (i.e. the complex conjugate of the ket wavefunction) and then integrate both sides of the equation over all space:

$$\langle\psi\mid O\mid\psi\rangle = \langle\psi\mid\lambda\mid\psi\rangle = \lambda\langle\psi|\psi\rangle = \lambda. \tag{1.22}$$

Assuming that our operator is Hermitian, λ is a real number and so can be factorised out of the integral. The remaining integral must equal 1 if we are working with normalised and orthogonal wavefunctions. Such wavefunctions are known as orthonormalised. In general wavefunctions that we encounter will be orthonormalised and operators will be Hermitian, so the above equation is widely applicable.

In some cases, when the operator acts on the ket wavefunction in the absence of the bra wavefunction (and so with no integration) it returns the wavefunction multiplied by a constant. In such cases, assuming the wavefunction is normalised, there is no need to proceed with the multiplication by the bra function and integrating the product, since the constant can be placed outside the integral and the integral will come to 1. The wavefunction may then simply be cancelled out, leaving the eigenvalue. Equation (1.12) is an example of this.

Chapter 2

Orbitals in the Hydrogen Atom

2.1 Spherical Polar Coordinates

In the hydrogen atom, it is easiest to consider the nucleus to be at the centre of our right-handed coordinate system. We assume that the nucleus is a stationary point. Given the spherical symmetry of an isolated hydrogen atom, it is most practical to use spherical polar coordinates. These are illustrated in Figure 2.1.

r is the radial distance, which is always positive. The angles θ and φ are a little like latitude and longitude when considering locations on the globe (except they are measured in degrees and latitude is measured from the equator). In spherical polar coordinates θ is known as the colatitude and lies in the range 0 to π radians; φ is known as the azimuth angle and lies in the range 0 to 2π radians. They are related to Cartesian coordinates as follows:

$$r = \sqrt{x^2 + y^2 + z^2} \tag{2.1a}$$

$$\theta = \arccos\left(z/\sqrt{x^2 + y^2 + z^2}\right) \tag{2.1b}$$

$$\varphi = \arctan\left(y/x\right). \tag{2.1c}$$

The inverse relations are

$$x = r \sin\theta \cos\varphi \tag{2.2a}$$

$$y = r \sin\theta \sin\varphi \tag{2.2b}$$

$$z = r \sin\theta. \tag{2.2c}$$

A great advantage of using spherical polar coordinates to describe the hydrogen electron is that the wavefunction can factorise into a radial part,

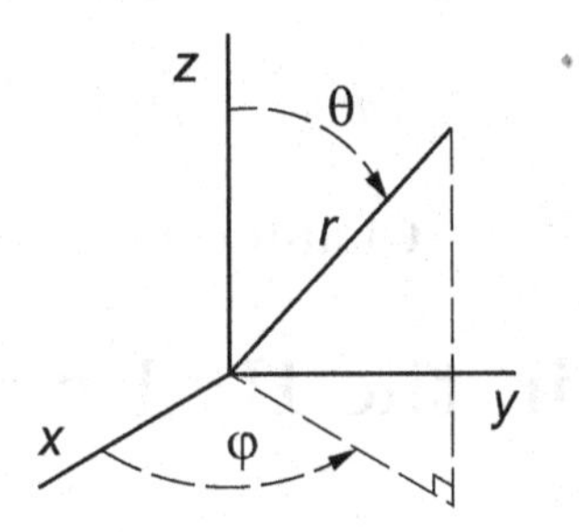

Figure 2.1. The spherical polar coordinate system.

i.e. depending only on r, and an angular part, depending only on θ and φ, which simplifies calculations.

2.2 Degrees of Freedom and Quantum Numbers

Since the earliest spectroscopic experiments in the nineteenth century it has been evident that atoms absorb and emit energy in discrete amounts. (The first equation to describe the energy gaps in hydrogen atoms using what was effectively a quantum number was devised by the Swiss-German schoolmaster J.J. Balmer in 1885.) This caused terrible problems for classical mechanics, which couldn't explain the results adequately.

Some simple analogies using standing waves can illustrate the origin of quantum numbers. Let us consider the standing wave on a plucked guitar string. The standing wave can only have certain wavelengths due to the constraint that the string is fixed at each end. In wave mechanics such a constraint is known as a boundary condition. The lowest energy (or fundamental) note is when the length of the string is half a wavelength, i.e. the only points on the string where the amplitude is fixed at zero are at the ends. These zero points are known as nodes. Higher energy standing waves have shorter wavelengths. The next-highest energy wave, the first harmonic, has half the wavelength of the fundamental, so that a whole wavelength is held by the string and there is a third node half way along the string. In music, these harmonic differences are the octaves. The harmonic to higher energy has one third of the wavelength of the fundamental. It turns out that the energy of the allowed standing waves is a function of $n - 1$, where n is the number of nodes on the string. ($n - 1$ is used so the fundamental takes a value of 1 and harmonics are multiples of this.) $n - 1$ here is analogous to a quantum number and its origin lies in the boundary condition imposed on the standing wave.[1]

In the plucked guitar string example, there is only one degree of freedom in the vibration as the string may be considered to be one-dimensional. A two-dimensional analogy would be a square drum in the xy-plane. When the membrane is struck, its vibrations have two degrees of freedom as the membrane can be considered two-dimensional. (The x- and y-axes are independent in that they are mutually perpendicular, or orthogonal.) This leads to two quantum numbers, n_x and n_y, say, where n_x is the number of nodes on the x-axis of the drum while n_y refers to the y-axis. Each quantum number relates to a boundary condition: the fact that the amplitude at the ends of the drum on each axis are fixed at zero. The energy of these standing waves would, analogously, be a function of $(n_x - 1)(n_y - 1)$. The three-dimensional analogue, a vibrating cube, is harder to visualise, but we could extend the analysis to conclude that it has three degrees of freedom, that its vibrations are associated with three quantum numbers, n_x, n_y and n_z, and that the energy of its standing waves is a function of $(n_x - 1)$ $(n_y - 1)(n_z - 1)$.

Analogies involving circular motion will bear a closer resemblance to the hydrogen atom. What about wavefunctions on a circular loop of wire? A circle may be thought to have no boundary but there is, in fact, a boundary condition. In order for a wave on the circular loop to be a standing wave, its amplitude at an angle φ from some reference point must be equal to its amplitude at the angle $\varphi + 2\pi$. While we may consider a circle in the xy-plane to be two-dimensional, if we describe it with spherical polar coordinates from the centre of the circle then r and θ are constant, leaving only a single coordinate, φ. So the circle is one-dimensional with one degree of freedom, and there is one boundary condition for its wavefunctions.

We could describe the wavefunction with a single quantum number, say m, that is the number of nodal planes passing through the standing wave on the loop (which will generate $2\,m$ nodal points on the function of the standing wave). Considering the wavefunctions as standing waves on a circle, successive harmonics have one additional nodal plane. It turns out that the energy of the circular wavefunctions is a function of m. By analogy with the harmonics on the one-dimensional string and equation (1.6), the

[1]It is conventional to number these principal-shell harmonics in atoms from 1, but other oscillating systems such as vibrations and rotations are numbered from zero.

general function for waves on a circular wire is

$$\psi = N_\varphi \mathrm{e}^{\mathrm{i}m\varphi}, \tag{2.3}$$

where N_φ is the arbitrary constant in front of any eigenfunction and m is the number of nodal planes associated with the standing wave. A linear combination of the m and $-m$ functions gives a real function for the standing wave with m nodal planes.

The two-dimensional analogy of a circular standing wave is a vibrating elastic sphere. In spherical polar coordinates measured from the centre of the sphere r is constant, leaving θ and φ as the two degrees of freedom, each with their own boundary conditions, $\psi(r,\theta,\varphi) = \psi(r,\theta+2\pi,\varphi)$ and $\psi(r,\theta,\varphi) = \psi(r,\theta,\varphi+2\pi)$.

There are two resulting quantum numbers, m relating to φ and l relating to θ. The nodes relating to l will be discussed further in Section 2.4.

The three-dimensional circular standing wave gives the solutions for the hydrogen electron. With r, θ and φ all variable there are three degrees of freedom, three boundary conditions and three quantum numbers. The boundary condition for r is that it must decay to zero at infinity — otherwise the electron wouldn't be localised on the atom. The quantum number associated with r is called n as there are certain parallels with the case of the vibrating string.

2.3 Angular Momentum

We saw in Section 1.3 that there is an operator for determining the momentum of an electron using an eigenvalue equation. This was a linear momentum, $m\mathbf{v}$ which is a vector since velocity ($\mathbf{v}$) is a vector. A more relevant quantity for circular motion is the angular momentum, $\mathbf{l}$, also a vector, which is defined as the cross product $\mathbf{r} \times \mathbf{p}$. If a particle is moving in a circle, the vector describing its angular momentum points perpendicular to the plane of the circle. Rather than defining the kinetic energy as $\frac{1}{2}mv^2 = p^2/2m$ we use the rotational kinetic energy $\frac{1}{2}I\omega^2 = l^2/2I$, where I is the moment of inertia of the electron and ω is the magnitude of the angular velocity (which is the linear velocity divided by the radius).

There are, however, three degrees of freedom to angular momentum, l_x, l_y and l_z. The quantum mechanics of angular momentum is a complicated subject because Heisenberg's uncertainty principle (which is a manifestation of the indeterminacy of certain quantities being measured simultaneously) forbids all three components of angular momentum being known simultaneously. The magnitude of the angular momentum and its component along a specified axis (l_z, by convention) can be known exactly.[2] The 3D angular momentum vector therefore has a known length and can be represented by the surface of a cone with known cone angle, but has unspecified direction.

l_z is defined as an anticlockwise rotation around the z-axis, going from the positive x-axis to the positive y-axis in our right-handed coordinate system, i.e. in the same sense that the angle φ is defined in Figure 2.1. Following the cross-product definition, and in accordance with an anti-clockwise rotation, we can express l_z as $xp_y - yp_x$. We convert this classical definition into a quantum mechanical operator as described in Section 1.3:

$$\hat{l_z} = \frac{\hbar}{\mathrm{i}}\left(x\frac{\partial}{\partial y} - y\frac{\partial}{\partial x}\right). \tag{2.4}$$

Descriptions of the angular momenta around all three axes can be generalised by a determinant,

$$\hat{\mathbf{L}} = \frac{\hbar}{\mathrm{i}}\begin{vmatrix} \mathbf{i} & \mathbf{j} & \mathbf{k} \\ x & y & z \\ \dfrac{\partial}{\partial x} & \dfrac{\partial}{\partial y} & \dfrac{\partial}{\partial z} \end{vmatrix}, \tag{2.5}$$

where $\mathbf{i}$, $\mathbf{j}$ and $\mathbf{k}$ are unit vectors pointing along the x-, y- and z-axes, respectively.

When this $\hat{l_z}$ angular momentum operator acts on an eigenfunction of angular momentum it returns the angular momentum projected on the z-axis, m, in units of $\hbar$:

$$l_z \mid \psi\rangle = \hbar m|\psi\rangle. \tag{2.6}$$

Exercise 5

Write down the operators for $\hat{l_x}$ and $\hat{l_y}$ in Cartesian coordinates, as in equation (2.4).

Exercise 6

Show that the operators for $\hat{l_x}$ and $\hat{l_y}$ do not commute by working out the commutator $[\hat{l_x}, \hat{l_y}]$.

[2]Strictly, angular momentum can never be known to arbitrary precision due to the indeterminacy associated with matter fields.

Exercise 7

Deduce the values of the other five commutators, $[\hat{l}_y, \hat{l}_x]$, $[\hat{l}_y, \hat{l}_z]$, etc.

Exercise 8

Use equations (2.1), (2.2) and (2.15) to work out $\hat{l}_x$, $\hat{l}_y$ and $\hat{l}_z$ in spherical polar coordinates.

Exercise 9

Show that $e^{im\varphi}$ is an eigenfunction of $\hat{l}_z$ and that equation (2.6) is true.

It is, however, permissible to know simultaneously l_z and the square magnitude of the total angular momentum, $l^2 = l_x^2 + l_y^2 + l_z^2$. This is because there is no directional information in l^2 since it is composed of scalar products from all three Cartesian axes. This permissibility can also be derived from the commutator of the relevant angular momentum operators. Since there are only two simultaneously observable orbital angular momenta for an electron, these are the quantities described by the two angular quantum numbers, l and m. (The definition of m used by chemists is equivalent to the m mentioned in Section 2.2 in the context of nodal planes.) l relates to the magnitude of the total orbital angular momentum of the electron in that state and so is always positive. It increases in integer steps in common with all the other quantum numbers we meet, with its minimum permissible value being 0 when there is no orbital angular momentum. m relates to the orientation of the angular momentum vector — specifically its projection on the z-axis. It also varies in integer steps, as discussed in Section 2.2. It is not surprising, then, that the permissible range of m is $-l, -l+1, \ldots, 0, 1, \ldots, l-1, l$. It follows that for each value of l there are $2l+1$ possible values of m.

Exercise 10

What is the total number of orbitals associated with a principal quantum shell, n?

2.4 Angular Wavefunctions of the Hydrogen Electron

The wavefunctions for an electron in a hydrogen atom apply generally to quantum particles in a centrosymmetric electrostatic field. An electron outside a core shell of shielding electrons in atom may also be described by similar functions.

Following the arguments in Sections 1.2 and 2.2, it is not surprising to find that the wavefunction involving φ, with its link to the one-dimensional circular motion and the m quantum number is the same as equation (2.3). The normalisation constant in front of each of the exponentials (see Section 1.5) is

$$N_\varphi = \frac{1}{\sqrt{2\pi}}. \tag{2.7}$$

The normalisation of this function is as follows. The complex conjugate of $N_\varphi \mathrm{e}^{\mathrm{i}m\varphi}$ is $N_\varphi \mathrm{e}^{-\mathrm{i}m\varphi}$. The product of these two functions is simply N_φ^2 since $\mathrm{e}^{\mathrm{i}m\varphi}\mathrm{e}^{-\mathrm{i}m\varphi} = \mathrm{e}^0 = 1$. N_φ is chosen so that the integral of $\psi^*\psi$ with respect to φ between 0 and 2π (which is all space for φ) gives 1. Since N_φ is just a number it is taken out of the integral:

$$N_\varphi^2 \int_0^{2\pi} 1\,\mathrm{d}\varphi = 1. \tag{2.8}$$

The integral comes to 2π so N_φ must equal $1/\sqrt{2\pi}$.

Much algebra is required to derive the functions of θ that are simultaneous eigenfunctions of l_z and l^2 (explained in [9]). They are the associated Legendre functions, which actually depend on $|m|$ as well as l. They are all real and given the symbol P_l^m. The P_l^0 functions can be found by constructing an orthogonal series of functions of increasing powers of $\cos\theta$, as described in Section 1.4. Ladder operators are required to generate the functions for different values of m [9]. The P_l^m for $l = 0$ to 3 are collected in Table 2.1 with the normalisation constants, N_θ [25], as discussed in Section 1.5.

Atomic orbitals can be described using spherical harmonic functions, $Y_l^m(\theta, \varphi)$. They are built from the associated Legendre functions, P_l^m, with their normalisation constant, N_θ, and the complex exponential of equation (2.3), $\mathrm{e}^{\mathrm{i}m\varphi}$, with its normalisation constant, N_φ.

Exercise 11

Show that the azimuthal wavefunctions $\mathrm{e}^{-\mathrm{i}m\varphi}/\sqrt{2\pi}$ and $\mathrm{e}^{\mathrm{i}m\varphi}/\sqrt{2\pi}$ are orthogonal by working out $\langle -m|m\rangle$.

Table 2.1. The associated Legendre functions, P_l^m, up to $l = 3$ with their normalisation constants, N_θ [25].

P_l^m	N_θ	$P_l^m(\theta)$
P_0^0	$\sqrt{2}/2$	1
P_1^0	$\sqrt{6}/2$	$\cos\theta$
$P_1^{\pm1}$	$\sqrt{3}/2$	$\sin\theta$
P_2^0	$\sqrt{10}/4$	$(3\cos^2\theta - 1)$
$P_2^{\pm1}$	$\sqrt{15}/2$	$\sin\theta\cos\theta$
$P_2^{\pm2}$	$\sqrt{15}/4$	$\sin^2\theta$
P_3^0	$\sqrt{14}/4$	$(5\cos^3\theta - 3\cos\theta)$
$P_3^{\pm1}$	$\sqrt{42}/8$	$\sin\theta(5\cos^2\theta - 1)$
$P_3^{\pm2}$	$\sqrt{105}/4$	$\sin^2\theta\cos\theta$
$P_3^{\pm3}$	$\sqrt{70}/8$	$\sin^3\theta$

Exercise 12

Show that the associated Legendre functions P_1^1 and P_1^0 are orthogonal by calculating $\langle P_1^1|P_1^0\rangle$.

Exercise 13

Show that the associated Legendre function P_2^0, the angular function for a d_{z^2} orbital, is correctly normalised.

The angular functions give the characteristic shapes of atomic orbitals. The l quantum number relates to the type of subshell, which is usually referred to by a letter, as shown in Table 2.2. The subshells taking values of l of 0, 1, 2 and 3 are given the labels s, p, d and f, respectively, for historical reasons. Considering s electrons classically, $\mathbf{r} \times \mathbf{p} = 0$, and so the position and momentum vectors must be pointing in the same direction, implying that s electron motion is back-and-forth through the nucleus rather than moving around it.

For non-zero m the $\psi(\varphi)$ are complex. These are made real by taking linear combinations, as shown in equation (1.8). The linear combinations are still solutions to the Hamiltonian and angular momentum eigenvalue equations, so this procedure is acceptable (see Section 1.4). Similarly, subtracting the complex conjugate function is acceptable since the resulting orbital is wholly imaginary. When this is multiplied by its complex conjugate to obtain the probability density function the result is wholly

Table 2.2. The orbital angular wavefunctions in real form for s, p, d and f orbitals with their normalisation constants, N_φ, and the linear combinations of the $e^{im\varphi}$ functions required [25]. For the form of the $P_l^m(\theta)$ functions see Table 2.1.

Orbital	N_θ	$P_l^m(\theta)$	Linear comb.	N_φ	$\psi(\varphi)$
s	$\sqrt{2}/2$	$P_0^0(\theta)$	$\|0\rangle$	$1/\sqrt{2\pi}$	1
p_z	$\sqrt{6}/2$	$P_1^0(\theta)$	$\|0\rangle$	$1/\sqrt{2\pi}$	1
p_x	$\sqrt{3}/2$	$P_1^1(\theta)$	$\|1\rangle + \|-1>$	$1/\sqrt{\pi}$	$\cos\varphi$
p_y	$\sqrt{3}/2$	$P_1^1(\theta)$	$\|1\rangle - \|-1>$	$1/\sqrt{\pi}$	$\sin\varphi$
d_{z^2}	$\sqrt{10}/4$	$P_2^0(\theta)$	$\|0\rangle$	$1/\sqrt{2\pi}$	1
d_{xz}	$\sqrt{15}/2$	$P_2^1(\theta)$	$\|1\rangle + \|-1>$	$1/\sqrt{\pi}$	$\cos\varphi$
d_{yz}	$\sqrt{15}/2$	$P_2^1(\theta)$	$\|1\rangle - \|-1>$	$1/\sqrt{\pi}$	$\sin\varphi$
$d_{x^2-y^2}$	$\sqrt{15}/4$	$P_2^2(\theta)$	$\|2\rangle + \|-2>$	$1/\sqrt{\pi}$	$\cos 2\varphi$
d_{xy}	$\sqrt{15}/4$	$P_2^2(\theta)$	$\|2\rangle - \|-2>$	$1/\sqrt{\pi}$	$\sin 2\varphi$
f_{z^3}	$\sqrt{14}/4$	$P_3^0(\theta)$	$\|0\rangle$	$1/\sqrt{2\pi}$	1
f_{xz^2}	$\sqrt{42}/8$	$P_3^1(\theta)$	$\|1\rangle + \|-1>$	$1/\sqrt{\pi}$	$\cos\varphi$
f_{yz^2}	$\sqrt{42}/8$	$P_3^1(\theta)$	$\|1\rangle - \|-1>$	$1/\sqrt{\pi}$	$\sin\varphi$
$f_{z(x^2-y^2)}$	$\sqrt{105}/4$	$P_3^2(\theta)$	$\|2\rangle + \|-2>$	$1/\sqrt{\pi}$	$\cos 2\varphi$
f_{xyz}	$\sqrt{105}/4$	$P_3^2(\theta)$	$\|2\rangle - \|-2>$	$1/\sqrt{\pi}$	$\sin 2\varphi$
$f_{x(x^2-3y^2)}$	$\sqrt{70}/8$	$P_3^3(\theta)$	$\|3\rangle + \|-3>$	$1/\sqrt{\pi}$	$\cos 3\varphi$
$f_{y(3x^2-y^2)}$	$\sqrt{70}/8$	$P_3^3(\theta)$	$\|3\rangle - \|-3>$	$1/\sqrt{\pi}$	$\sin 3\varphi$

real and positive. Note that the angular momentum l_z of all the real linear combinations effectively disappears: the angular momentum vector must therefore lie in the xy-plane, causing the electrons to rotate in a vertical cycle. However the real functions are no longer strictly eigenfunctions of angular momentum (though they are of $\hat{\mathbf{l}}^2$). Real sine and cosine functions cannot be eigenfunctions of a momentum operator since the operator involves a first derivative (e.g. equation (1.12)), which returns a different trigonometric function. They may however be eigenfunctions of the kinetic energy or Hamiltonian operators since these involve second derivatives (equation (2.22)), returning the same trigonometric function. The real orbital functions together with their chemical labels are collected in Table 2.2.

Exercise 14

Show that the linear combination $(e^{im\varphi} + e^{-im\varphi})/\sqrt{4\pi}$ is orthogonal to $(e^{im\varphi} - e^{-im\varphi})/\sqrt{4\pi}$ by calculating the integral $1/\sqrt{4\pi} \times \langle e^{im\varphi} + e^{-im\varphi}|e^{im\varphi} - e^{-im\varphi}\rangle$.

Table 2.3. The orbital angular functions for s, p, d and f orbitals in their full Cartesian form with normalisation constants. See the text for an explanation of the conversion.

Orbital	Normalisation	Full Cartesian function
s	$1/(2\sqrt{\pi})$	1
p_z	$\sqrt{3}/(2\sqrt{\pi})$	z/r
p_x	$\sqrt{3}/(2\sqrt{\pi})$	x/r
p_y	$\sqrt{3}/(2\sqrt{\pi})$	y/r
d_{z^2}	$\sqrt{5}/(4\sqrt{\pi})$	$(3z^2 - r^2)/r^2$
d_{xz}	$\sqrt{15}/(2\sqrt{\pi})$	xz/r^2
d_{yz}	$\sqrt{15}/(2\sqrt{\pi})$	yz/r^2
$\mathrm{d}_{x^2-y^2}$	$\sqrt{15}/(4\sqrt{\pi})$	$(x^2 - y^2)/r^2$
d_{xy}	$\sqrt{15}/(2\sqrt{\pi})$	xy/r^2
f_{z^3}	$\sqrt{7}/(4\sqrt{\pi})$	$z(5z^2 - 3r^2)/r^3$
f_{xz^2}	$\sqrt{42}/(8\sqrt{\pi})$	$x(5z^2 - r^2)/r^3$
f_{yz^2}	$\sqrt{42}/(8\sqrt{\pi})$	$y(5z^2 - r^2)/r^3$
$\mathrm{f}_{z(x^2-y^2)}$	$\sqrt{105}/(4\sqrt{\pi})$	$z(x^2 - y^2)/r^3$
f_{xyz}	$\sqrt{105}/(2\sqrt{\pi})$	xyz/r^3
$\mathrm{f}_{x(x^2-3y^2)}$	$\sqrt{70}/(8\sqrt{\pi})$	$x(x^2 - 3y^2)/r^3$
$\mathrm{f}_{y(3x^2-y^2)}$	$\sqrt{70}/(8\sqrt{\pi})$	$y(3x^2 - y^2)/r^3$

The Cartesian labels for the real forms of the orbitals can be obtained by converting the $\psi(\theta, \varphi)$ to $\psi(x, y, z)$ using equations (2.2) and the first of equations (2.1). As well as the well-known double-angle formulae, $\cos 2\varphi = \cos^2\varphi - \sin^2\varphi$ and $\sin 2\varphi = 2\cos\varphi \sin\varphi$, we need the higher multiple-angle trigonometric identities in equations (2.9) to establish the Cartesian labels for the f and higher orbitals. The full Cartesian expressions are collected in Table 2.3.

$$\cos 3\varphi = \cos\varphi(\cos^2\varphi - 3\sin^2\varphi) \tag{2.9a}$$

$$\sin 3\varphi = \sin\varphi(3\cos^2\varphi - \sin^2\varphi). \tag{2.9b}$$

When linear combinations are taken of complex exponentials to make orbital functions real (or wholly imaginary) the normalisation of the resultant trigonometric functions must be considered. These functions will be $2\cos m\varphi$ and $2\mathrm{i}\sin m\varphi$. The 2 (or indeed any constant) in front of these functions may be disregarded as the normalisation process provides the constant by which the function has to be multiplied for the wavefunction to be valid.[3] Hence Dirac describes eigenvectors as vectors which are defined by the direction they are pointing in phase space and whose length is just a result of normalisation [7]. All the functions $\cos m\varphi$ and $\mathrm{i}\sin m\varphi$ are normalised by $1/\sqrt{\pi}$. When either of these functions is multiplied by its complex conjugate it returns its real square, which is converted to a simple trigonometric form using the double-angle cosine formula:

$$N_\varphi^2 \int_0^{2\pi} \cos^2 m\varphi \, \mathrm{d}\varphi = N_\varphi^2 \int_0^{2\pi} \left(\frac{1}{2} + \frac{1}{2}\cos 2m\varphi\right) \mathrm{d}\varphi = 1. \tag{2.10}$$

Integration leads to the following expression:

$$N_\varphi^2 \left[\frac{\varphi}{2} + \frac{1}{4m}\sin 2m\varphi\right]_0^{2\pi} = N_\varphi^2 \pi = 1. \tag{2.11}$$

Solving the above expression for N_φ gives the normalisation constant as predicted. A similar process is followed for $i\sin m\varphi$:

$$N_\varphi^2 \int_0^{2\pi} \sin^2 m\varphi \, \mathrm{d}\varphi = N_\varphi^2 \int_0^{2\pi} \left(\frac{1}{2} - \frac{1}{2}\cos 2m\varphi\right) \mathrm{d}\varphi = 1. \tag{2.12}$$

Figures 2.2, 2.3, 2.4 and 2.5 show the exact forms of the amplitude for the angular wavefunctions of the s, p, d and f orbitals, respectively. Where the amplitude is negative, the orbital is depicted with broken lines. They appear to be three-dimensional plots, but are only two-dimensional in terms of spherical polar coordinates, since they are just functions of θ and φ. The plots are not isosurfaces as the radial part of the wavefunction is not included.

The nodes in the orbital angular functions are planes or cones. The number of nodal planes or cones is equal to l, i.e. no nodal plane or cone

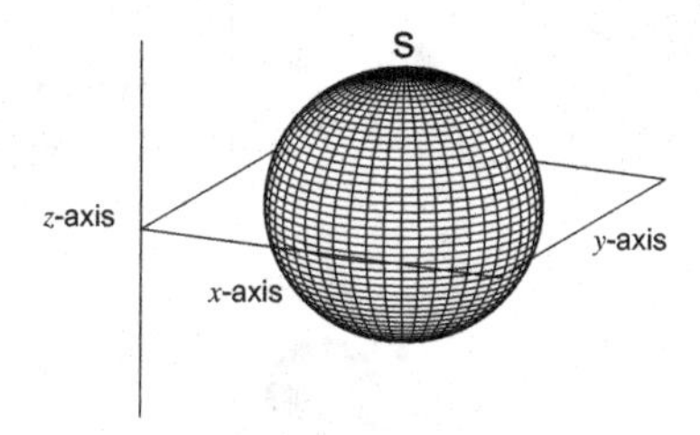

Figure 2.2. The angular wavefunction for the atomic s orbital.

[3]The i may also be disregarded as it has no physical meaning.

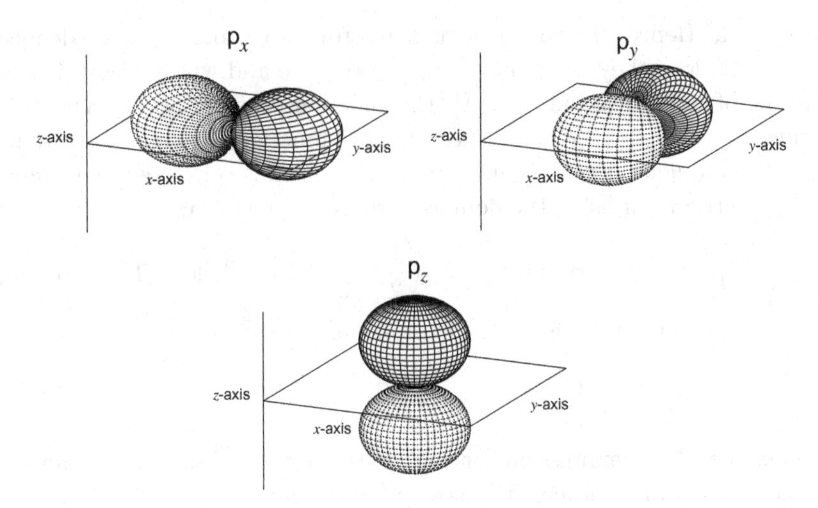

Figure 2.3. The angular wavefunctions for the atomic p orbitals. The broken lines indicate negative phase.

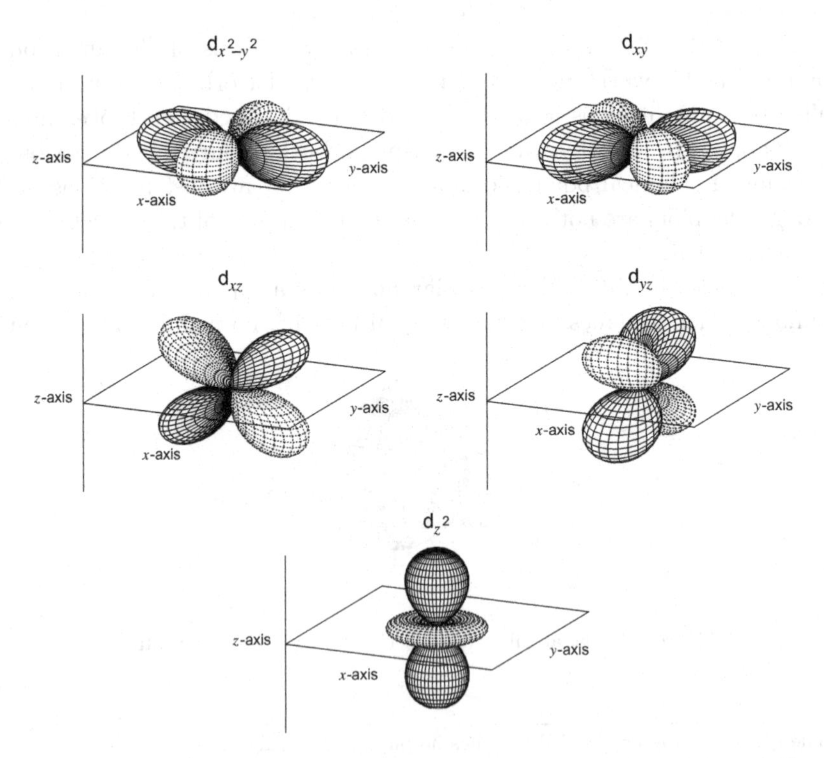

Figure 2.4. The angular wavefunctions for the atomic d orbitals. The broken lines indicate negative phase.

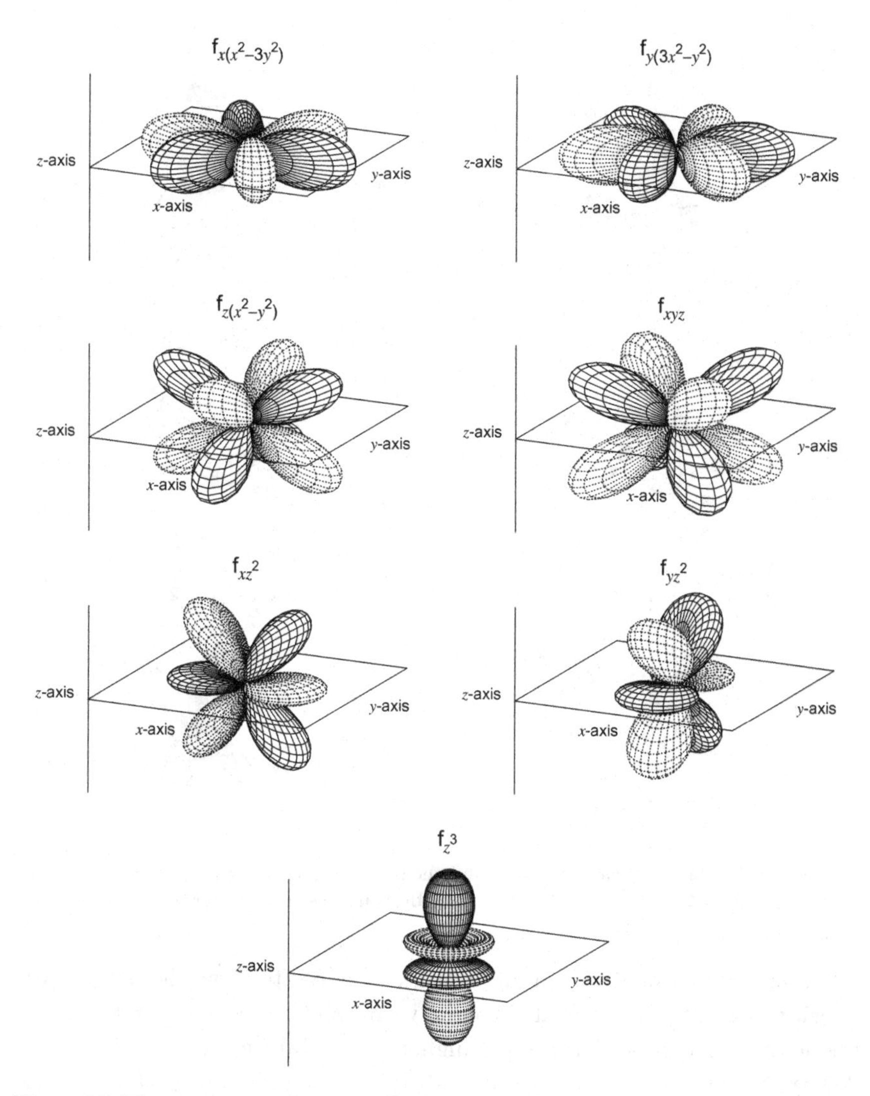

Figure 2.5. The angular wavefunctions for the atomic f orbitals. The broken lines indicate negative phase.

for s orbitals, one for p orbitals (the yz-plane for the p_x orbital, etc.) and two for d orbitals (the xz- and yz-planes for the d_{xy} orbital, etc.). In the case of d_{z^2} there are two nodal cones, one above and one below the xy-plane. When $m = 0$, $e^{im\varphi} = 1$ and so in these cases the angular function is just the appropriately normalised associated Legendre function.

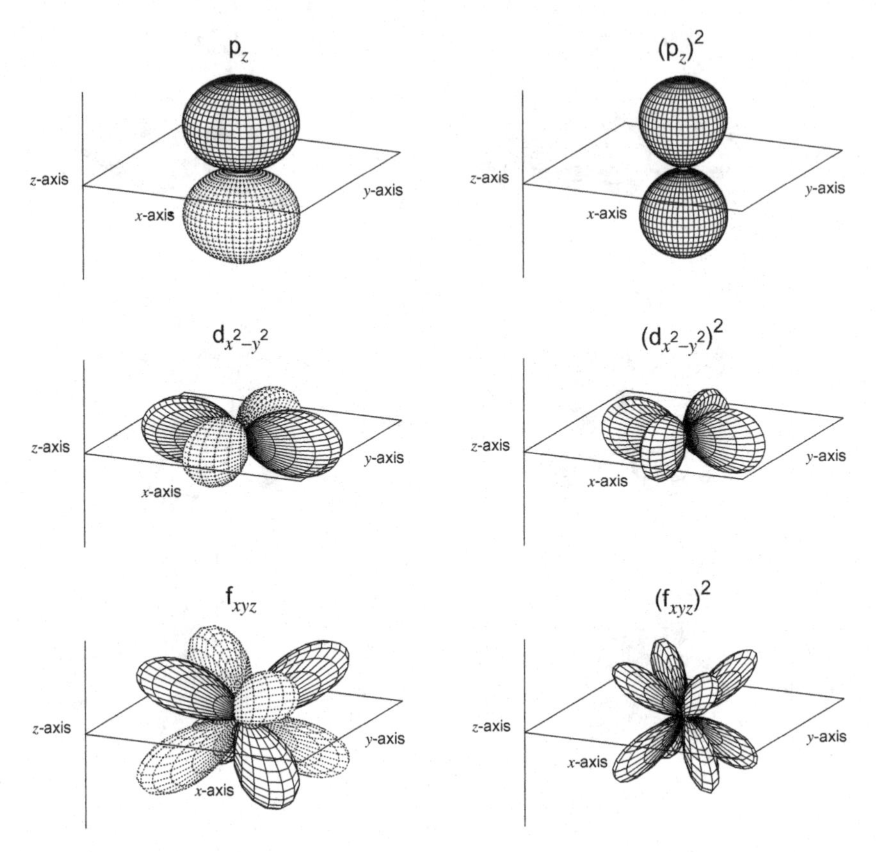

Figure 2.6. The angular wavefunctions and their squares, i.e. their probability densities, for the atomic p_z, $d_{x^2-y^2}$, f_{xyz} orbitals. The broken lines indicate negative phase.

It should be noted that if an orbital is squared to show the probability density then its shape will change, with lobes appearing slimmer, as trigonometric functions raised to higher powers are more tightly curved. There are, of course, no phases associated with probability densities, as probability is always positive. Figure 2.6 shows p_z, $d_{x^2-y^2}$ and f_{xyz} orbitals alongside their probability densities to illustrate the difference. In each subshell the value of l gives the power to which trigonometric functions are raised; this power is doubled in the probability density.

It has been widely commented upon that among the d orbitals, d_{z^2} appears to be the odd one out. While this is unavoidable in spherical symmetry, it has been pointed that the five d orbitals have the same appearance in an environment with five-fold symmetry [26].

Some generalisations can be made about the shapes of these orbitals. Since the orbitals are defined with respect to a unique z-axis, this is the axis about which every orbital has most rotational symmetry. The $|m|$ value associated with each orbital or linear combination therefore gives information about the power to which z is raised in the Cartesian form of the angular probability density functions. Specifically, z is raised to the power of $l - |m|$.

Within a given subshell the number of lobes on an orbital depends on the number of Cartesian axes that appear in the orbital label. In any subshell, when $m = 0$ there is only one Cartesian axis, the z-axis, in the orbital label, raised to the power of l. These orbitals always have $l + 1$ lobes, counting each cylindrically symmetrical ring as a lobe. In the context of the most commonly seen subshells, s, p and d, the $m = 0$ d orbital is indeed the odd one out, having a different number of lobes to the other orbitals in its subshell. However, when considering higher subshells we realise that in fact the s and p subshells are special cases in having all their orbitals with the same number of lobes in spherical symmetry.

2.5 Cubic f Orbitals in the Hydrogen Atom

It is common to work with atomic orbitals in cubic symmetry, i.e. assuming that the x-, y- and z-axes are equivalent, as is the case around tetrahedral and octahedral centres. The real orbitals defined earlier for the s, p and d subshells already have cubic symmetry (despite the unavoidably different appearance of the d_{z^2} orbital). The same is not true for the f subshell where it is clear that the orbitals have been defined with respect to a unique (z) axis.

The f orbitals may be transformed into cubic sets by examining how their angular functions behave under the rotations of the octahedral group. Since these rotations interconvert x, y and z it is easiest to work with the hydrogen angular wavefunctions in Cartesian form [27, 28]. Orbitals are grouped together that are either invariant under rotations or are rotated into one another. Linear combinations of orbitals that behave similarly under the rotations are then orthonormalised, i.e. their coefficients are chosen so that when the product of one orbital's wavefunction with the complex conjugate of another is integrated over all space the result is 0 apart from when the two orbitals are the same, in which case the result is 1.

Table 2.4. The orbital angular wavefunctions for f orbitals in cubic symmetry in terms of spherical harmonic functions [28].

ψ(cubic)	$\psi(Y_3^m(\theta,\varphi))$
$\|\mathrm{f\ A_2\ a_2}\rangle$	$\frac{1}{\sqrt{2}}Y_3^2 - \frac{1}{\sqrt{2}}Y_3^{-2}$
$\|\mathrm{f\ T_1\ 1}\rangle$	$-\frac{\sqrt{5}}{2\sqrt{2}}Y_3^{-3} - \frac{\sqrt{3}}{2\sqrt{2}}Y_3^1$
$\|\mathrm{f\ T_1\ 0}\rangle$	Y_3^0
$\|\mathrm{f\ T_1 - 1}\rangle$	$-\frac{\sqrt{5}}{2\sqrt{2}}Y_3^3 - \frac{\sqrt{3}}{2\sqrt{2}}Y_3^{-1}$
$\|\mathrm{f\ T_2\ 1}\rangle$	$-\frac{\sqrt{3}}{2\sqrt{2}}Y_3^3 + \frac{\sqrt{5}}{2\sqrt{2}}Y_3^{-1}$
$\|\mathrm{f\ T_2\ 0}\rangle$	$\frac{1}{\sqrt{2}}Y_3^2 + \frac{1}{\sqrt{2}}Y_3^{-2}$
$\|\mathrm{f\ T_2 - 1}\rangle$	$-\frac{\sqrt{3}}{2\sqrt{2}}Y_3^{-3} + \frac{\sqrt{5}}{2\sqrt{2}}Y_3^1$

The cubic linear combinations of hydrogen angular wavefunctions for the f subshell in terms of their spherical harmonic functions are collected in Table 2.4 using the symmetry labels of Griffith [28]. The spherical harmonic functions $Y_l^m(\theta,\varphi)$ are defined as

$$Y_l^m(\theta,\varphi) = (-1)^{m>0} N P_l^m(\theta) \mathrm{e}^{\mathrm{i}m\varphi}, \tag{2.13}$$

where N is a normalisation constant that can be found from Table 2.2. The strange-looking phase factor is intended to convey the approach taken by from Condon and Shortley in their classic treatise [29], which is used by most authors. The spherical harmonic function only takes a negative phase when m is both odd and positive. This is a natural choice when the spherical harmonic functions are generated by a lowering operator [9].

Some of the cubic angular wavefunctions shown in Table 2.4 mix $|3\ m\rangle$ eigenfunctions with different magnitudes of m. These functions remain complex and so the linear combinations shown in Table 2.5 are required for all the cubic orbitals to be real. Note that the $m = 0$ function is entirely real. The signs are chosen so that the resulting angular orbital functions are positive.

It is evident from Table 2.4 that some of the f orbitals are the same in both spherical and octahedral symmetry. The f orbitals in these two symmetries are shown in Table 2.6. The spherical labels denote the m labels used in linear combinations: $\pm m$ denotes the linear combination $|m\rangle + |-m\rangle$ and $\mp m$ denotes the linear combination $|m\rangle - |-m\rangle$. The octahedral label

Table 2.5. Linear combinations of the cubic angular wavefunctions in the f subshell from Table 2.4 required to generate real functions, following the phases adopted by Griffith [28].

Cubic f orbital	Normalisation	Linear combination
f_{x^3}	$-(1/\sqrt{2})$	$\lvert f\, T_1\, 1\rangle - \lvert f\, T_1 - 1\rangle$
f_{y^3}	$i(1/\sqrt{2})$	$\lvert f\, T_1\, 1\rangle + \lvert f\, T_1 - 1\rangle$
$f_{x(z^2-y^2)}$	$(1/\sqrt{2})$	$\lvert f\, T_2\, 1\rangle - \lvert f\, T_2 - 1\rangle$
$f_{y(z^2-x^2)}$	$i(1/\sqrt{2})$	$\lvert f\, T_2\, 1\rangle + \lvert f\, T_2 - 1\rangle$

Table 2.6. The spherical and cubic f orbitals. See the text for the explanation of the labels.

Spherical f orbital	Label		Cubic f orbital	Label
f_{z^3}	0	=	f_{z^3}	T_1
f_{xz^2}	± 1		f_{x^3}	T_1
f_{yz^2}	∓ 1		f_{y^3}	T_1
$f_{z(x^2-y^2)}$	± 2	=	$f_{z(x^2-y^2)}$	T_2
f_{xyz}	∓ 2	=	f_{xyz}	A_2
$f_{x(x^2-3y^2)}$	± 3		$f_{x(z^2-y^2)}$	T_2
$f_{y(3x^2-y^2)}$	∓ 3		$f_{y(z^2-x^2)}$	T_2

Table 2.7. The orbital angular functions for cubic f orbitals in their full Cartesian form with normalisation constants.

f orbital	Normalisation	Full Cartesian function
f_{z^3}	$\sqrt{7}/(4\sqrt{\pi})$	$z(5z^2-3r^2)/r^3$
f_{x^3}	$\sqrt{7}/(4\sqrt{\pi})$	$x(5x^2-3r^2)/r^3$
f_{y^3}	$\sqrt{7}/(4\sqrt{\pi})$	$y(5y^2-3r^2)/r^3$
$f_{z(x^2-y^2)}$	$\sqrt{105}/(4\sqrt{\pi})$	$z(x^2-y^2)/r^3$
f_{xyz}	$\sqrt{105}/(2\sqrt{\pi})$	xyz/r^3
$f_{x(z^2-y^2)}$	$\sqrt{105}/(4\sqrt{\pi})$	$x(z^2-y^2)/r^3$
$f_{y(z^2-x^2)}$	$\sqrt{105}/(4\sqrt{\pi})$	$y(z^2-x^2)/r^3$

classifies the orbitals by how they are transformed by octahedral symmetry operations (two-, three- and four-fold rotations, etc.).

The cubic f orbital angular functions in Cartesian form are collected in Table 2.7. The functions are calculated using the entries in Tables 2.2, 2.3, 2.4 and 2.5.

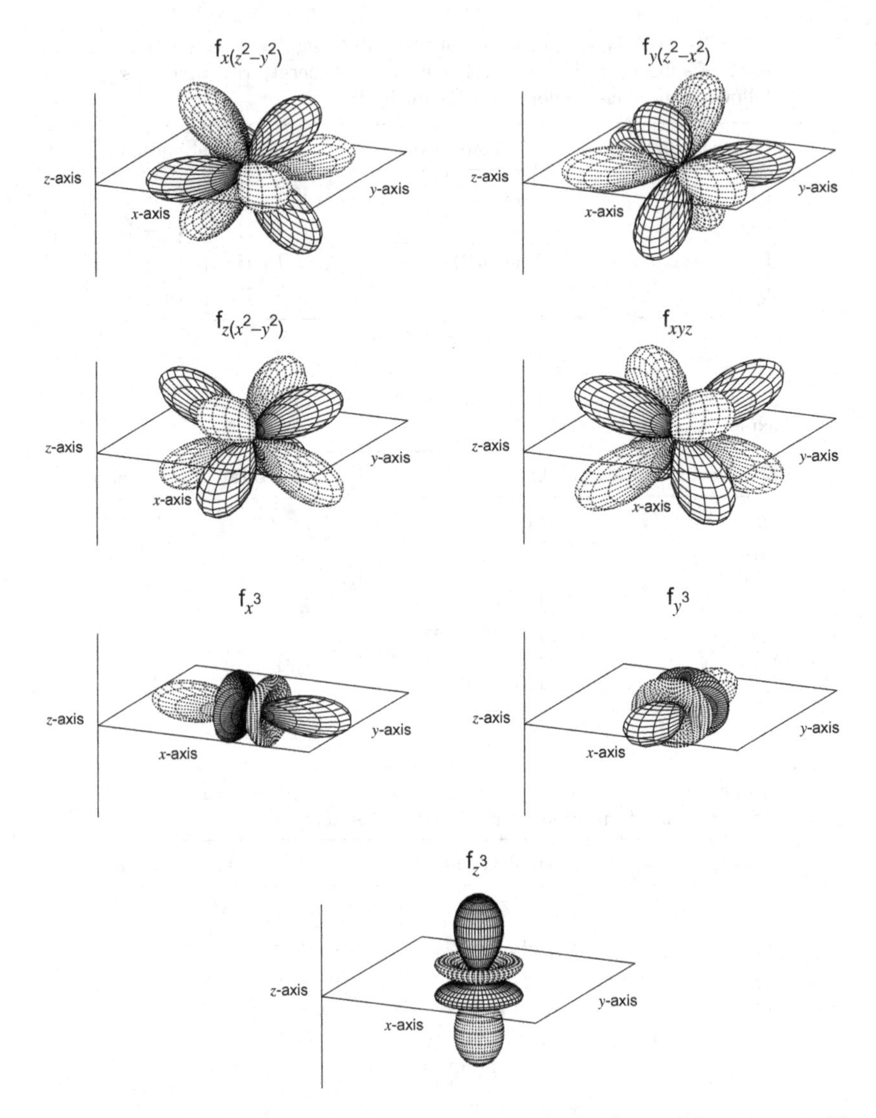

Figure 2.7. The angular wavefunctions for the atomic f orbitals in cubic symmetry. The broken lines indicate negative phase.

The cubic f orbital angular wavefunctions are plotted in Figure 2.7. Where the amplitude is negative the orbital is depicted with broken lines.

As was seen with the spherical f orbitals, the number of lobes seen in a cubic orbital within a given subshell depends on the number of different Cartesian axes appearing in the orbital's label. Inspection of these labels reveals the equal status of the Cartesian axes, and this symmetry is of course reflected in the appearance of the orbitals.

2.6 Radial Wavefunctions of the Hydrogen Electron

We consider now the Schrödinger equation for the hydrogen electron. As mentioned in Section 1.3 it is an eigenvalue equation. The operator is known as the Hamiltonian, given the symbol $\hat{H}$, and is the operator for total energy, i.e. kinetic energy + potential energy. It therefore returns the total energy of the electron as its eigenvalue. Using the quantum operator for momentum and for distance we can translate the classical expression for total energy,

$$E = \frac{1}{2m}\left(p_x^2 + p_y^2 + p_z^2\right) - \frac{Ze^2}{4\pi\varepsilon_0\sqrt{x^2+y^2+z^2}}, \tag{2.14}$$

into a wave mechanical equation. In equation (2.14) the second energy term is the potential energy of the electron. It is the electrostatic potential energy and is negative because of the attraction between the electron and the hydrogen nucleus. e is the fundamental charge and ε_0 is the permittivity of free space. Z is the number of protons in the nucleus. This is useful because ions like He^+ and Li^{2+} are 'hydrogen-like' in that they only have one electron.

As explained in Section 2.1 it is most convenient to express the wavefunction in spherical polar coordinates. The Hamiltonian operator will therefore need to be transformed so that it acts on spherical polar coordinates. This is achieved for functions using equations (2.2) but is more complicated for the differential operators required for the momentum operator. The theory of partial derivatives is required and we make use of the following well-known result for differentiation:

$$\frac{\partial\psi}{\partial x} = \left(\frac{\partial\psi}{\partial r}\times\frac{\partial r}{\partial x}\right) + \left(\frac{\partial\psi}{\partial\theta}\times\frac{\partial\theta}{\partial x}\right) + \left(\frac{\partial\psi}{\partial\varphi}\times\frac{\partial\varphi}{\partial x}\right). \tag{2.15}$$

The transformation is more complicated for the second derivative. The second derivative with respect to x is shown below just in terms of the spherical polar coordinate r:

$$\begin{aligned}\frac{\partial^2\psi}{\partial x^2} &= \frac{\partial}{\partial x}\left(\frac{\partial\psi}{\partial x}\right)\\ &= \left(\frac{\partial\psi}{\partial r}\times\frac{\partial^2 r}{\partial x^2}\right) + \frac{\partial r}{\partial x}\left(\frac{\partial^2\psi}{\partial r^2}\times\frac{\partial r}{\partial x}\right)\\ &\quad + \text{similar terms involving } \theta \text{ and } \varphi.\end{aligned} \tag{2.16}$$

Equations (2.15) and (2.16) transform the second derivative with respect to x to spherical polar operators. Together with similar expressions for $\frac{\partial^2\psi}{\partial y^2}$ and $\frac{\partial^2\psi}{\partial z^2}$, the required second derivative with respect to all space may be found. The symbol ∇, pronounced 'del' or 'nabla', is used to denote differentiation with respect to all space. The second derivative with respect to all space, ∇^2, is known as the Laplacian:

$$\nabla^2 = \frac{\partial^2}{\partial x^2} + \frac{\partial^2}{\partial y^2} + \frac{\partial^2}{\partial z^2}. \tag{2.17}$$

Expressions such as $\frac{\partial r}{\partial x}$ are obtained from equations (2.1). Completing this lengthy algebra gives the Laplacian in spherical polar coordinates [9].

$$\nabla^2 = \frac{1}{r^2}\left\{\frac{\partial}{\partial r}\left(r^2\frac{\partial}{\partial r}\right) + \left(\frac{1}{\sin\theta}\frac{\partial}{\partial\theta}\sin\theta\frac{\partial}{\partial\theta}\right) + \left(\frac{1}{\sin^2\theta}\frac{\partial^2}{\partial\varphi^2}\right)\right\}. \tag{2.18}$$

Exercise 15

Work out $\partial^2 r/\partial x^2$, $\partial^2\theta/\partial x^2$ and $\partial^2\varphi/\partial x^2$ and use these results to find $\partial^2\psi/\partial x^2$.

Exercise 16

Following a similar method to the last exercise find $\partial^2\psi/\partial y^2$ and $\partial^2\psi/\partial z^2$ and use these to derive the result in equation (2.18).

While on the subject of transforming between coordinate systems it is also worth noting that when integrating over all space, $\mathrm{d}x\,\mathrm{d}y\,\mathrm{d}z$, the equivalent in spherical polar coordinates is found using the ratio of the volume elements, which is calculated from the Jacobian determinant of the transformation [9]. The result is shown below:

$$\begin{aligned}\langle\psi|\psi\rangle &= \int_{-\infty}^{\infty}\int_{-\infty}^{\infty}\int_{-\infty}^{\infty}\psi^*(x,y,z)\psi(x,y,z)\,\mathrm{d}x\,\mathrm{d}y\,\mathrm{d}z \\ &= \int_0^{\infty}\int_0^{2\pi}\int_0^{\pi}\psi^*(r,\theta,\varphi)\psi(r,\theta,\varphi)\,r^2\sin\theta\,\mathrm{d}\theta\,\mathrm{d}\varphi\,\mathrm{d}r. \end{aligned} \tag{2.19}$$

Exercise 17

The Jacobian matrix transforms the partial differentials $(\partial r\ \partial\theta\ \partial\varphi)$ into $(\partial x\ \partial y\ \partial z)$, encapsulating equation (2.15) in matrix form:

$$\begin{bmatrix} \frac{\partial x}{\partial r} & \frac{\partial x}{\partial \theta} & \frac{\partial x}{\partial \varphi} \\ \frac{\partial y}{\partial r} & \frac{\partial y}{\partial \theta} & \frac{\partial y}{\partial \varphi} \\ \frac{\partial z}{\partial r} & \frac{\partial z}{\partial \theta} & \frac{\partial z}{\partial \varphi} \end{bmatrix} \begin{bmatrix} \partial r \\ \partial \theta \\ \partial \varphi \end{bmatrix} = \begin{bmatrix} \partial x \\ \partial y \\ \partial z \end{bmatrix}. \tag{2.20}$$

The determinant of a matrix gives the scale factor involved in the transformation. Use this information to derive the result in equation (2.19).

A significant point about the Laplacian operator, which is part of the momentum operator (and therefore the kinetic energy operator) is that the final two terms in equation (2.18), i.e. the functions of θ and φ, are equal to the operator $-\hat{l}^2$ in spherical polar coordinates, which gives the square magnitude of the orbital angular momentum of the electron (see Section 2.3). When this operates on the wavefunction, the quantum number for the magnitude of the orbital angular momentum, l, is produced as part of the eigenvalue. As a result, the radial wavefunctions will depend on l as well as any other quantum number that arises from a boundary condition. The radial wavefunctions are, however, independent of the m quantum number.

Exercise 18

Use the relation $\hat{l}^2 = \hat{l}_x^2 + \hat{l}_y^2 + \hat{l}_z^2$ to find $\hat{l}^2$ in spherical polar coordinates.

The spherical harmonic functions, or suitably normalised products of the associated Legendre functions and the complex exponentials $\mathrm{e}^{\mathrm{i}m\varphi}$, are eigenfunctions of the $\hat{l}^2$ operator, with eigenvalues of $\hbar^2 l(l+1)$. For a proof, see reference [9].

$$\hat{l}^2\left(P_l^m \mathrm{e}^{\mathrm{i}m\varphi}\right) = \hbar^2 l(l+1)\left(P_l^m \mathrm{e}^{\mathrm{i}m\varphi}\right). \tag{2.21}$$

Exercise 19

Show that equation (2.21) gives the expected eigenvalue when $\hat{l}^2$ acts on $\sin^2\theta\cos 2\varphi$, the $\mathrm{d}_{x^2-y^2}$ angular function.

It is a long and difficult process to find the radial eigenfunction solutions to the Schrödinger equation below for the hydrogen electron. Interested readers should consult one of the advanced texts in the bibliography [9,10,25].

$$\left\{\frac{-\hbar^2}{2\mu r^2}\left(\frac{\partial}{\partial r}r^2\frac{\partial}{\partial r}+\frac{1}{\sin\theta}\frac{\partial}{\partial\theta}\sin\theta\frac{\partial}{\partial\theta}+\frac{1}{\sin^2\theta}\frac{\partial^2}{\partial\varphi^2}\right)-\frac{Ze^2K}{r}\right\}\psi=E\psi, \tag{2.22}$$

where $\hbar$ is $h/2\pi$. μ is the reduced mass of the electron, which takes into account the fact that the electron and nucleus both rotate around a common centre of mass (rather than the nucleus being at the centre of the rotation of the electron).

$$\mu=\frac{m_e.m_N}{m_e+m_N}, \tag{2.23}$$

where m_e is the mass of the electron and m_N is the mass of the nucleus. μ provides only a slight correction to the mass of the electron since its mass is so small compared with the mass of the nucleus. K is shorthand for the constants in the Coulomb expression:

$$K=\frac{1}{4\pi\varepsilon_0}. \tag{2.24}$$

The minus sign in front of the kinetic energy operator in equation (2.22) is the result of applying the momentum operator twice, which also accounts for the second derivative. Since the momentum operator contains the imaginary i, applying it twice gives $\mathrm{i}^2=-1$.

Following the earlier discussion of the Laplacian operator it is evident that, considering the second and third terms in equation (2.22) as one, we can describe the three terms on the left-hand side of the equation as the operators for linear kinetic energy, rotational kinetic energy and potential energy, respectively.

The crucial point in the algebra of finding the radial eigenfunctions is setting the boundary condition that the radial function should decay to zero as r approaches infinity. As with the examples in Section 2.2 the boundary condition leads to a quantum number, which is the principal (shell) quantum number n, which can take values 1, 2, 3,.... The full analysis [9] yields the following expression for n:

$$n=\frac{Ze^2K}{\hbar}\sqrt{\frac{\mu}{-2E}}, \tag{2.25}$$

where E is the total energy of the electron, which is defined to be negative (zero at the ionisation limit). It can also be shown that for a given value of n, l may take values in the range 0, 1, ..., $n-1$ [9].

The radial eigenfunctions turn out to be the associated Laguerre functions, which depend on the quantum numbers n and l, and therefore pertain to each subshell. They are shown in Table 2.8 with their normalisation constants, N_r.

In Table 2.8, a_0 is the Bohr radius, about 52.9 pm (1 pm = 10^{-12} m), which is the most probable distance of the electron from the nucleus in hydrogen. It can be expressed as a series of constants (see Section 2.7):

$$a_0 = \frac{\hbar^2}{m_e e^2 K}. \tag{2.26}$$

Table 2.8. The radial wavefunctions for the subshells of the first three shells, with their normalisation constants, N_r [25]. Note that ρ is defined as in equation (2.27), differently to most sources (see below).

Subshell	N_r	$\psi(r)$
1s	1	$(Z/a_0)^{3/2}.2\mathrm{e}^{-\rho r/2}$
2s	$1/(2\sqrt{2})$	$(Z/a_0)^{3/2}.(2-\rho r)\mathrm{e}^{-\rho r/2}$
2p	$1/(2\sqrt{6})$	$(Z/a_0)^{3/2}.\rho r\mathrm{e}^{-\rho r/2}$
3s	$1/(9\sqrt{3})$	$(Z/a_0)^{3/2}.(6-6\rho r+\rho^2r^2)\mathrm{e}^{-\rho r/2}$
3p	$1/(9\sqrt{6})$	$(Z/a_0)^{3/2}.(4-\rho r)\rho r\mathrm{e}^{-\rho r/2}$
3d	$1/(9\sqrt{30})$	$(Z/a_0)^{3/2}.\rho^2r^2\mathrm{e}^{-\rho r/2}$
4s	$1/96$	$(Z/a_0)^{3/2}.(24-36\rho r+12\rho^2r^2-\rho^3r^3)\mathrm{e}^{-\rho r/2}$
4p	$1/(32\sqrt{15})$	$(Z/a_0)^{3/2}.(20-10\rho r+\rho^2r^2)\rho r\mathrm{e}^{-\rho r/2}$
4d	$1/(96\sqrt{5})$	$(Z/a_0)^{3/2}.(6-\rho r)\rho^2r^2\mathrm{e}^{-\rho r/2}$
4f	$1/(96\sqrt{35})$	$(Z/a_0)^{3/2}.\rho^3r^3\mathrm{e}^{-\rho r/2}$
5s	$1/(300\sqrt{5})$	$(Z/a_0)^{3/2}.(120-240\rho r+120\rho^2r^2-20\rho^3r^3$ $+\rho^4r^4)\mathrm{e}^{-\rho r/2}$
5p	$1/(150\sqrt{30})$	$(Z/a_0)^{3/2}.(120-90\rho r+18\rho^2r^2-\rho^3r^3)\rho r\mathrm{e}^{-\rho r/2}$
5d	$1/(150\sqrt{70})$	$(Z/a_0)^{3/2}.(42-14\rho r+\rho^2r^2)\rho^2r^2\mathrm{e}^{-\rho r/2}$
5f	$1/(300\sqrt{70})$	$(Z/a_0)^{3/2}.(8-\rho r)\rho^3r^3\mathrm{e}^{-\rho r/2}$
5g	$1/(900\sqrt{70})$	$(Z/a_0)^{3/2}.\rho^4r^4\mathrm{e}^{-\rho r/2}$
6s	$1/(2160\sqrt{6})$	$(Z/a_0)^{3/2}.(720-1800\rho r+1200\rho^2r^2-300\rho^3r^3$ $+30\rho^4r^4-\rho^5r^5)\mathrm{e}^{-\rho r/2}$
6p	$1/(432\sqrt{210})$	$(Z/a_0)^{3/2}.(840-840\rho r+252\rho^2r^2-28\rho^3r^3$ $+\rho^4r^4)\rho r\mathrm{e}^{-\rho r/2}$

where m_e is the mass of the electron. In very accurate work, the Bohr radius will be modified slightly to take into account the reduced mass of the electron. From the definition of K in equation (2.24) and the electrostatic energy expression, the base units of K are kg m^3 s^{-2} C^{-2}. Given that the base units of h are kg m^2 s^{-1}, one can appreciate that equation (2.26) is dimensionally correct. The constant ρ that appears in the radial wavefunctions is a collection of constants that has dimensions of reciprocal distance:

$$\rho = \frac{2Z}{na_0}. \tag{2.27}$$

Note that ρ is defined here so as not to include the radial variable r, which is different from how it is defined in most texts. The reason is to make r explicit in the wavefunctions as, being radial wavefunctions, they are nominally functions of r.

With dimensions of reciprocal distance, ρ ensures that the exponents and the terms involving r before the exponential function are dimensionless. This means that all the radial functions have dimensions of (distance)$^{-3/2}$. This is required for them to be normalised since the integral over all space of the square modulus of the wavefunction must equal 1. Following equation (2.19), when a function of purely radial coordinates is integrated over all space, the Jacobian determinant introduces r^2 so that the integral takes the following form:

$$\int_0^\infty \psi(r)^2\, r^2\, \mathrm{d}r = 1. \tag{2.28}$$

Exercise 20

Show that the 1s radial wavefunction in Table 2.8 is correctly normalised.

The square of the wavefunction multiplied by r^2 gives a function of inverse distance which, integrated over r, leads to a pure number, which is what is required for a probability. The normalised radial wavefunctions for the subshells of the first three shells are plotted in Figure 2.8.

Exercise 21

Show that the 1s wavefunction in Table 2.8 is a solution to the Schrödinger equation (equation (2.22)) and that the expression for its energy (equation (2.33)) is valid.

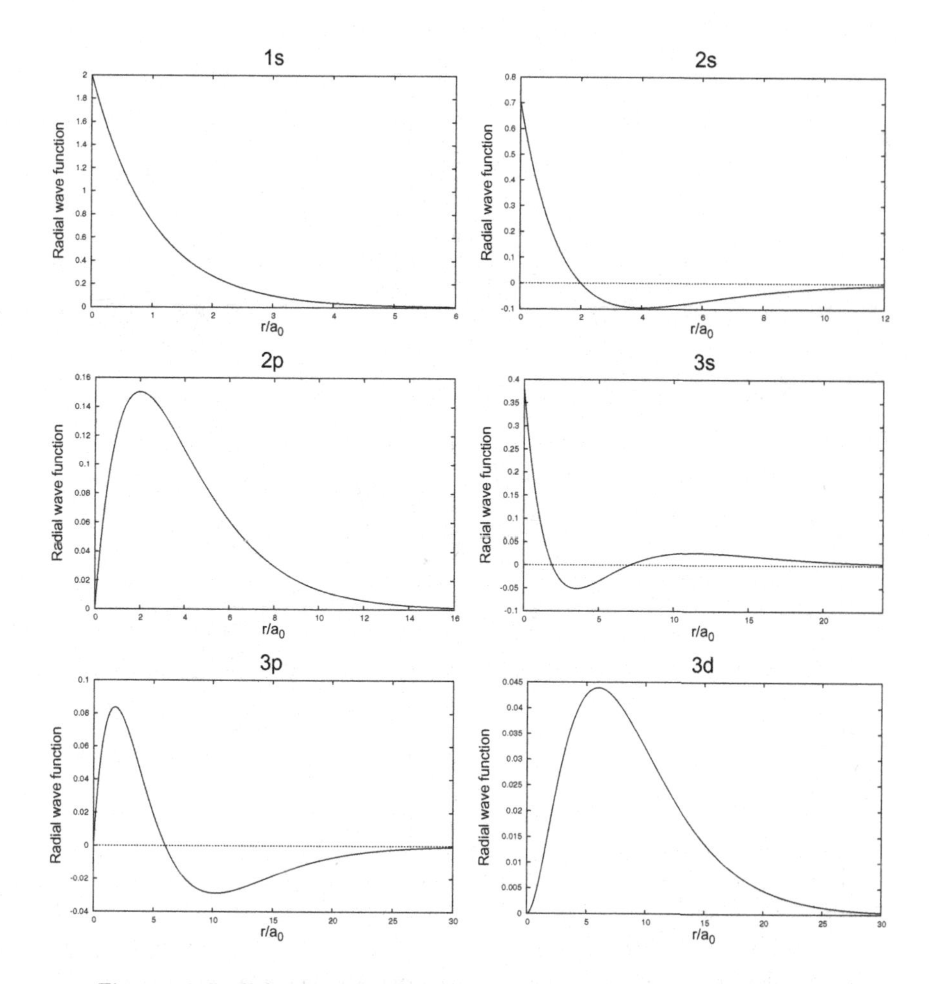

Figure 2.8. Radial wavefunctions for H atom orbitals of the first three shells.

Exercise 22

Show that the 1s and 2s radial wavefunctions in Table 2.8 are orthogonal to each other.

The normalised radial wavefunctions for six higher subshells from the fourth, fifth and sixth shells are plotted in Figure 2.9. Note that these are the orbitals for hydrogen ($Z = 1$), which accounts for the very large radial extent of these higher orbitals.

A common way to visualise radial wavefunctions is the radial density function (RDF). The RDF describes the probability of finding an

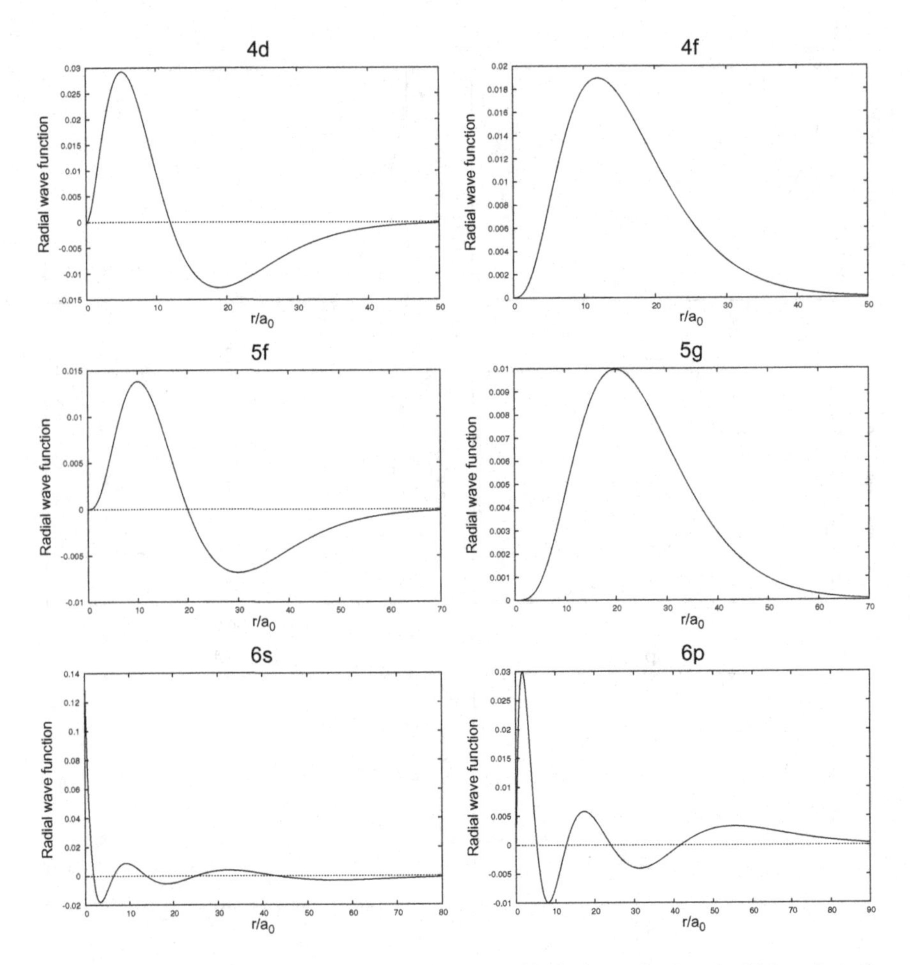

Figure 2.9. Radial wavefunctions for six H atom orbitals from the fourth, fifth and sixth shells.

electron a distance r from the nucleus, i.e. in the shell $4\pi r^2 \delta r$. Following equation (2.28) the RDF is defined as $4\pi r^2 \psi(r)^2$, which integrates over all r to give 1. The radial density functions for the subshells of the first three shells are plotted in Figure 2.10.

Exercise 23

Show that the Bohr radius is the most probable distance of a 1s electron from a hydrogen nucleus.

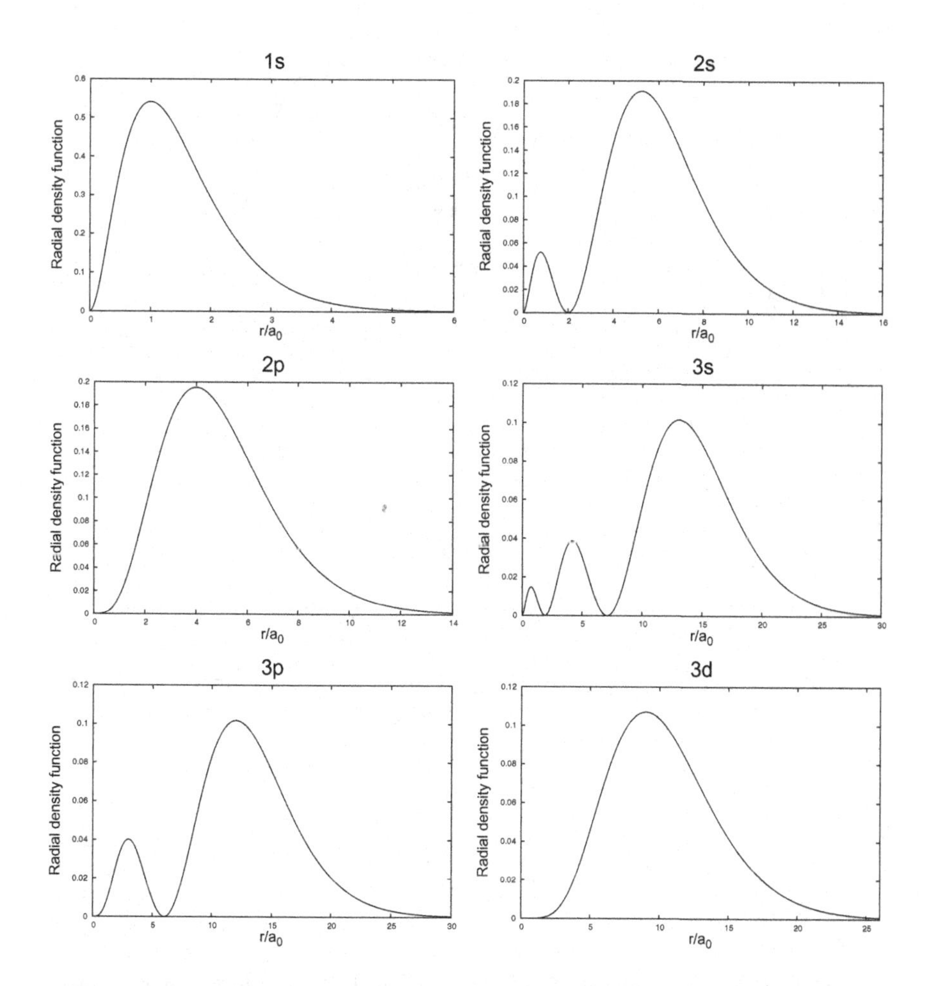

Figure 2.10. Radial density functions for H atom subshells of the first three shells.

The hydrogenic radial density functions for the six higher subshells from the fourth, fifth and sixth shells in Figure 2.9 are plotted in Figure 2.11.

Inspection of Figures 2.8 to 2.11 shows that the number of nodes shown by the radial density function, excluding $r = 0$ and ∞, is given by $n - l - 1$. These radial nodes are distinct from the angular nodes discussed in Section 2.4.

The radial and angular nodes are both visible in the probability density map of the xy-plane for the 2s and $3d_{xy}$ orbitals in Figure 2.12. Being a probability density, all points are non-negative. The 2s orbital just has one

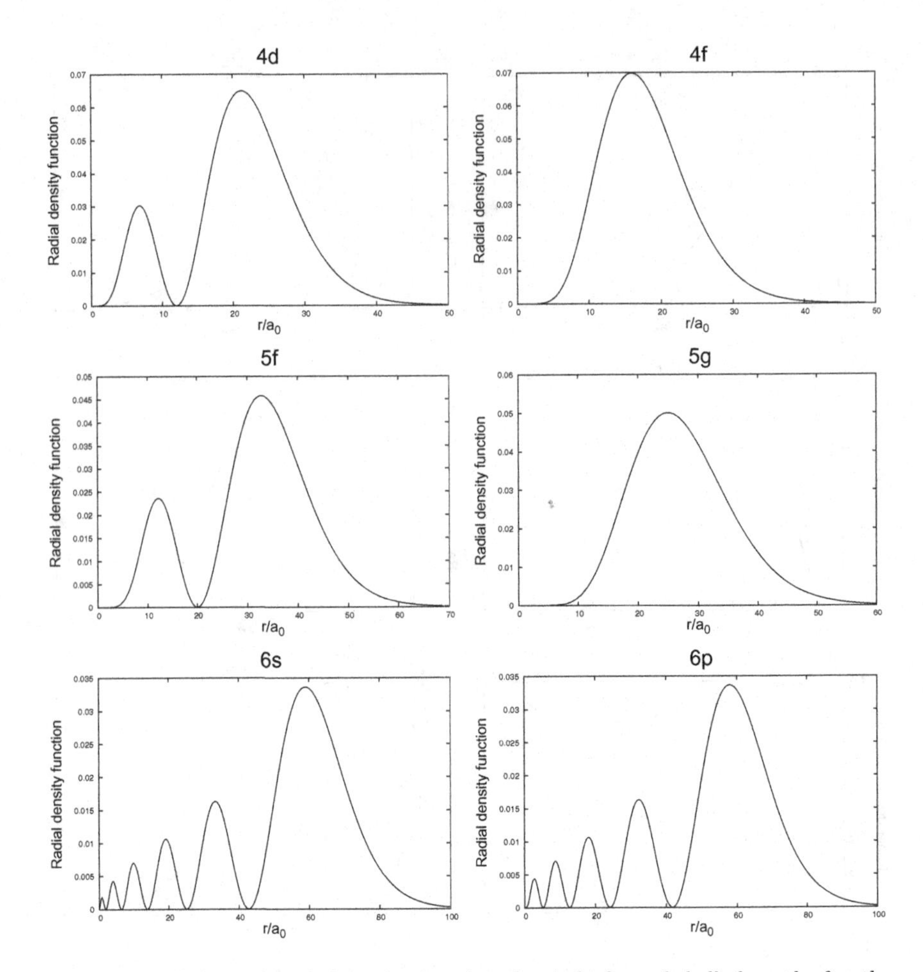

Figure 2.11. Hydrogenic radial density functions for six higher subshells from the fourth, fifth and sixth shells.

radial node at the bottom of the spike. The $3d_{xy}$ orbital has no radial node but two angular nodal planes (xz- and yz-planes).

Figure 2.13 gives the contour maps in the xy-plane of the wavefunction amplitudes for a 2s and a $3p_x$ orbital. Negative values are plotted with broken lines. The 2s contour map reveals a radial node at $r = 2$ Bohr radii; for $3p_x$ it reveals both an angular and a radial node. The angular nodal plane is the yz-plane, and the radial node is at about $r = 5$ Bohr radii.

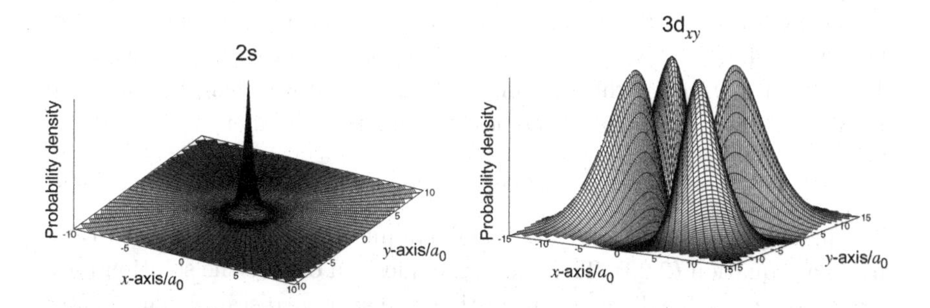

Figure 2.12. Probability density for the H atom $2s$ and $3d_{xy}$ orbitals in the xy-plane.

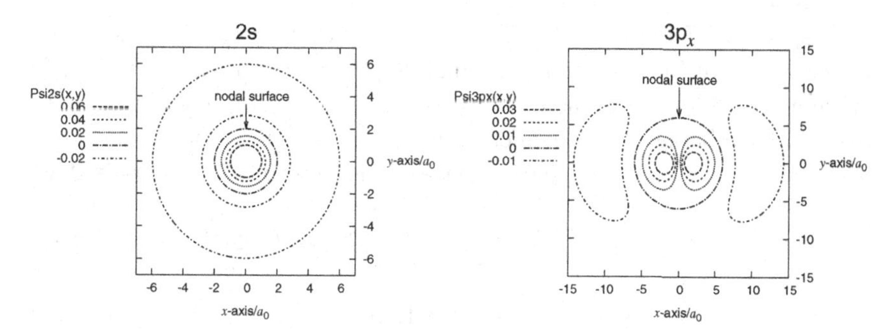

Figure 2.13. Wavefunction amplitude contour maps for the H atom $2s$ and $3p_x$ orbitals in the xy-plane.

Exercise 24

Use the 2s orbital radial wavefunction in Table 2.8 to show that the 2s orbital has a radial node at $r = 2a_0$.

2.7 The Bohr Radius

A remarkable fact about the expression for the Bohr radius in equation (2.26) is that it can be derived in a simple semi-classical way. If we picture an electron orbiting a proton such that the electrostatic force of attraction is equal to the centripetal force, then

$$\frac{e^2K}{r^2} = \frac{m_e v^2}{r}. \tag{2.29}$$

We then quantise angular momentum, using a quantum number, n, and $\hbar$. This is justified by considering the electron motion around the nucleus in the form of a wave in which an integral number of wavelengths makes up one revolution. This integer becomes the quantum number n:

$$m_e v r = n\hbar. \tag{2.30}$$

By equating equations (2.29) and (2.30), eliminating v and setting n to 1, we arrive at equation (2.26). This raises the question of why the solution for r is the most probable distance rather than the average distance. The answer is justified by Louis de Broglie in his discussion of least action (action has the units of momentum × distance, like $\hbar$) in part 5 of the first chapter of his classic book [30].

Inspection of Figures 2.10 and 2.11 shows that the subshell with the highest value of l in a shell has a radial density function with just a single maximum. It is therefore these functions that are easiest to use to generalise the most probable electron–nuclear distance as a function of the shell number. These maximum-l radial density functions may be generalised as

$$\mathrm{RDF}_{l_{\max}}(n) = N^2 \left(\frac{2Z}{na_0}\right)^{2(n-1)} r^{2n} \exp\left(\frac{-2Zr}{na_0}\right), \tag{2.31}$$

where N is the normalisation constant in front of the wavefunction. As these functions only have a single maximum, the most probable electron–nuclear distance is found by differentiating the radial density function with respect to r, setting it to zero and solving for r. This gives the most probable distance, r_{mp}, as

$$r_{\mathrm{mp}}(n) = \frac{n^2 a_0}{Z}. \tag{2.32}$$

This is, of course, consistent with the definition of the Bohr radius as the most probable distance of the hydrogen electron from the nucleus in its ground state. For these maximum-l orbitals the most probable distance of the hydrogen electron from the nucleus therefore increases rapidly with n, which can seem surprising given the knowledge that for high n the electron energy in hydrogen converges to a limit. Inspection of Figures 2.10 and 2.11 shows that the most probable distance of the hydrogen electron from the nucleus in the 2p, 3d, 4f and 5g orbitals in units of a_0 is 4, 9, 16 and 25, respectively. It is interesting that the most probable distance expression is the inverse of the expectation value for $1/r$. This again is in accord with the least action argument of de Broglie.

Exercise 25

Differentiate equation (2.31) with respect to r and show that equation (2.32) is correct.

2.8 Orbital Energies in the Hydrogen Atom

Rearrangement of the expression for n (equation (2.25)) gives the energy of the electron as a function of the quantum number n, which is typically simplified by substituting in the expression for the Bohr radius, a_0 (equation (2.26)):

$$E_n = -\frac{m_e Z^2 e^4 K^2}{2\hbar^2 n^2} = -\frac{Z^2 e^2 K}{2n^2 a_0}. \tag{2.33}$$

The energy of the electron is defined to be negative since it is trapped in the attractive potential well of the nuclear attraction. This equation may be simplified using the Rydberg constant for hydrogen, R_H:

$$E_n = -\frac{Z^2 R_H}{n^2}, \tag{2.34}$$

where R_H takes a value of 13.6 eV, 2.18×10^{-18} J or 1312 kJ mol^{-1}. After the fourth significant figure it differs from the more general Rydberg constant, R_∞, used in physics, as the Rydberg constant for hydrogen uses the reduced mass of the electron orbiting the proton rather than its rest mass.

Exercise 26

Calculate the energy in kJ mol^{-1} of the first excited state of the hydrogen atom above the ground state, and the ionisation energy of the hydrogen atom.

Exercise 27

Light was detected at 1.21 nm from a recent supernova due to the $n = 2$ to $n = 1$ transition on a hydrogenic atom from a heavier element. Work out which element was the source of the light.

Not just the Bohr radius, but the energy of the hydrogen electron can also be determined semi-classically. The sum of the classical kinetic and electrostatic potential energies from the circular motion of a negatively charged particle around a much heavier positively charged particle is

$$E = \frac{m_e v^2}{2} - \frac{Ze^2 K}{r}. \tag{2.35}$$

Eliminating r from equations (2.29) and (2.30) gives the following expression for v:

$$v = \frac{Ze^2 K}{n\hbar}. \tag{2.36}$$

Similarly v can be eliminated from equations (2.29) and (2.30) to give the following expression for r, which can be simplified by substituting in the expression for the Bohr radius (equation (2.26)):

$$r = \frac{n^2 \hbar^2}{Ze^2 K m_e} = \frac{n^2 a_0}{Z}. \tag{2.37}$$

Substituting equations (2.36) and (2.37) into (2.35) gives the following expression for the energy of a hydrogen electron as a function of the quantum number n:

$$E(n) = \frac{Z^2 e^2 K}{2n^2 a_0} - \frac{Z^2 e^2 K}{n^2 a_0} = -\frac{Z^2 e^2 K}{2n^2 a_0}. \tag{2.38}$$

This expression makes it clear that the potential energy (the second term) takes a value of $-2\times$ the kinetic energy (making the total energy equal to minus the kinetic energy), in accordance with the virial theorem (equation (2.41)).

It is interesting to note that the total energy of the electron in a one-electron atom depends only on the quantum number n, not the subshell quantum number l. That is to say that all orbitals and subshells within a given principal shell in the hydrogen atom are degenerate, i.e. of the same energy. This raises a question. Is the ratio of kinetic to potential energy the same for electrons in the same shell but different subshell? Electrons with a greater l value have greater angular momentum (since l is the quantum number for the magnitude of the orbital angular momentum) so they must possess greater rotational kinetic energy.

We can find the average value of observable quantities that are inherently probabilistic by working our their expectation values. In general the expectation value, $\langle \lambda \rangle$, of an observable quantity, λ, with the quantum operator $\hat{O}$ is given by

$$\langle \lambda \rangle = \langle \psi \,|O|\, \psi \rangle\,. \tag{2.39}$$

The potential energy operator is $1/r$ multiplied by a constant, $-e^2/4\pi\varepsilon_0$ or $-e^2K$. A very useful result is the expectation value for the operator $1/r$ when applied to the associated Laguerre functions, i.e. $\psi(r)$ [10]:

$$\left\langle \frac{1}{r} \right\rangle = \left\langle \psi(r) \left| \frac{1}{r} \right| \psi(r) \right\rangle = \frac{Z}{n^2 a_0}. \tag{2.40}$$

Exercise 28

Show that equation (2.40) is true for the 1s radial function.

Exercise 29

Show that equation (2.40) is consistent with equation (2.33) for the total energy of the hydrogen electron.

The significance of this result is that the potential energy of an electron in a one-electron atom depends only on the shell quantum number n, not the angular momentum (subshell) quantum number l. A very general result known as the virial theorem, which applies to classical as well as quantum systems, imposes that the kinetic energy, T, and the potential energy, V, for system of conservative forces, i.e. one where energy isn't exchanged with the environment, are related.[4] In the case of electrostatic attraction, the relation is

$$\langle T \rangle = -\frac{1}{2}\langle V \rangle. \tag{2.41}$$

Knowledge of the potential energy of the system consequently leads directly to the kinetic energy and therefore the total energy, which must equal $\langle V \rangle/2$. Taking the expectation value for the $1/r$ operator in equation (2.40), dividing it by 2 and multiplying by the constant $-e^2K$ gives the total energy of the electron given in equation (2.33).

Another useful result is the expectation value $\langle r \rangle$ for the hydrogen electron. Its value for the associated Laguerre functions is [10]

$$\langle r \rangle = \frac{n^2 a_0}{Z}\left\{1 + \frac{1}{2}\left(1 - \frac{l(l+1)}{n^2}\right)\right\}, \tag{2.42}$$

which shows that $\langle r \rangle$ decreases as l increases. This is evident from inspection of the radial density functions in Figures 2.10 and 2.11. While s orbitals

[4]The virial theorem applies to gravitational as well as electrostatic forces, so it holds on the scale of stars and planetary systems as well as atoms.

have the greatest $\langle r \rangle$ in a given shell one can appreciate how the $1/r$ is equal to other subshells since there is greater probability density for s electrons close to the nucleus due to larger number of radial nodes.

It is possible to find an expression for the variation of $\langle r \rangle$ as a function of n most easily when considering the s orbital in each shell, as the final term in equation (2.42) will disappear. This gives $\langle r_s \rangle(n)$ as

$$\langle r_s \rangle(n) = \frac{3n^2 a_0}{2Z}, \tag{2.43}$$

which, interestingly, is 1.5 times the expression for $r_{mp}(n)$ in equation (2.32).

So if the potential energy of the hydrogen electron is independent of the l quantum number then the same must be true of the kinetic energy. Since the angular momentum of the hydrogen electron does depend on l, then electrons with higher l must have greater rotational kinetic energy. Therefore they must have lower linear kinetic energy so that their total kinetic energy is constant.

Exercise 30

Use equation (2.22) to show that the three 2p orbitals in an isolated hydrogen atom all have the same energy.

Exercise 31

Use equations (2.22) and (2.40) to show that an electron occupying a 2s orbital in hydrogen has the same energy as an electron in a 2p orbital.

In equation (2.18) the final two terms in the operator that involve derivatives with respect to angles are part of the rotational kinetic energy operator, while the term involving derivatives with respect to r is part of the linear kinetic energy operator. The value of l corresponds to the power to which the trigonometric function of θ is raised in the wavefunction. The second derivative of l relates to the curvature of the trigonometric function of θ, which increases as l increases. s orbitals have no trigonometric function of θ as $l = 0$, but they have more radial nodes $(n - l - 1)$ than orbitals with higher values of l. More nodes means greater curvature of the radial wavefunction, and the second derivative of r relates to the curvature of the radial function, which explains why s orbitals have greater linear kinetic energy than other orbitals in a given shell.

2.9 g Orbitals

Chemists are familiar with s, p, d and f orbitals, which are occupied in the ground state configuration of the heavier elements in the periodic table. There is, however, no limit to the value that can be taken, in principle, by l (higher subshells can be detected spectroscopically in excited states). There has been a recent claim for the discovery of an element with an atomic number of about 122 [5]. The aufbau principle (see Section 3.2) would predict the ground-state occupation of an $l = 4$ subshell for such an element, which is given the label g. This g orbital assignment is consistent with recent relativistic calculations [6]. For this reason, g orbitals will be considered now.

It should be pointed out, however, that in any atom heavy enough to have an occupied g subshell in its ground-state configuration there will be significant relativistic effects such as spin–orbit coupling (see Section 4.5). Under these conditions it would actually be more realistic to consider spin-dependent orbitals. The g orbitals presented in this chapter are relevant for excited states of hydrogen, though (or Rydberg excited states of light atoms).

The g orbital angular wavefunctions are plotted in Figure 2.14. Where the amplitude is negative the orbital is depicted with broken lines.

The same approach for s, p, d and f orbitals is extended to g orbitals. Associated Legendre functions with $l = 4$ are required, which are given in Table 2.9. They are combined with real combinations of $e^{im\varphi}$, given in Table 2.10.

Cartesian forms of the wavefunctions are given in Table 2.11. They are found by applying the multiple-angle formulae of equations (2.44), together with the earlier identities, and substituting in the Cartesian relations of equations (2.2):

$$\cos 4\varphi = \cos^4 \varphi + \sin^4 \varphi - 6\cos^2 \varphi \sin^2 \varphi) \tag{2.44a}$$

$$\sin 4\varphi = 4(\cos^3 \varphi \sin \varphi - \cos \varphi \sin^3 \varphi). \tag{2.44b}$$

A cubic set of g wavefunctions may be created using similar methods to those employed for the f orbitals (see Section 2.5). Table 2.12 gives the linear combination of (complex) spherical harmonics (defined in equation (2.13)) that give cubic wavefunctions, and the symmetry labels that describe how they transform under octahedral symmetry operations. Five of these nine linear combinations are real because they involve $m = 0$ or linear combinations of $|m\rangle$ and $|-m\rangle$. The remaining four linear combinations

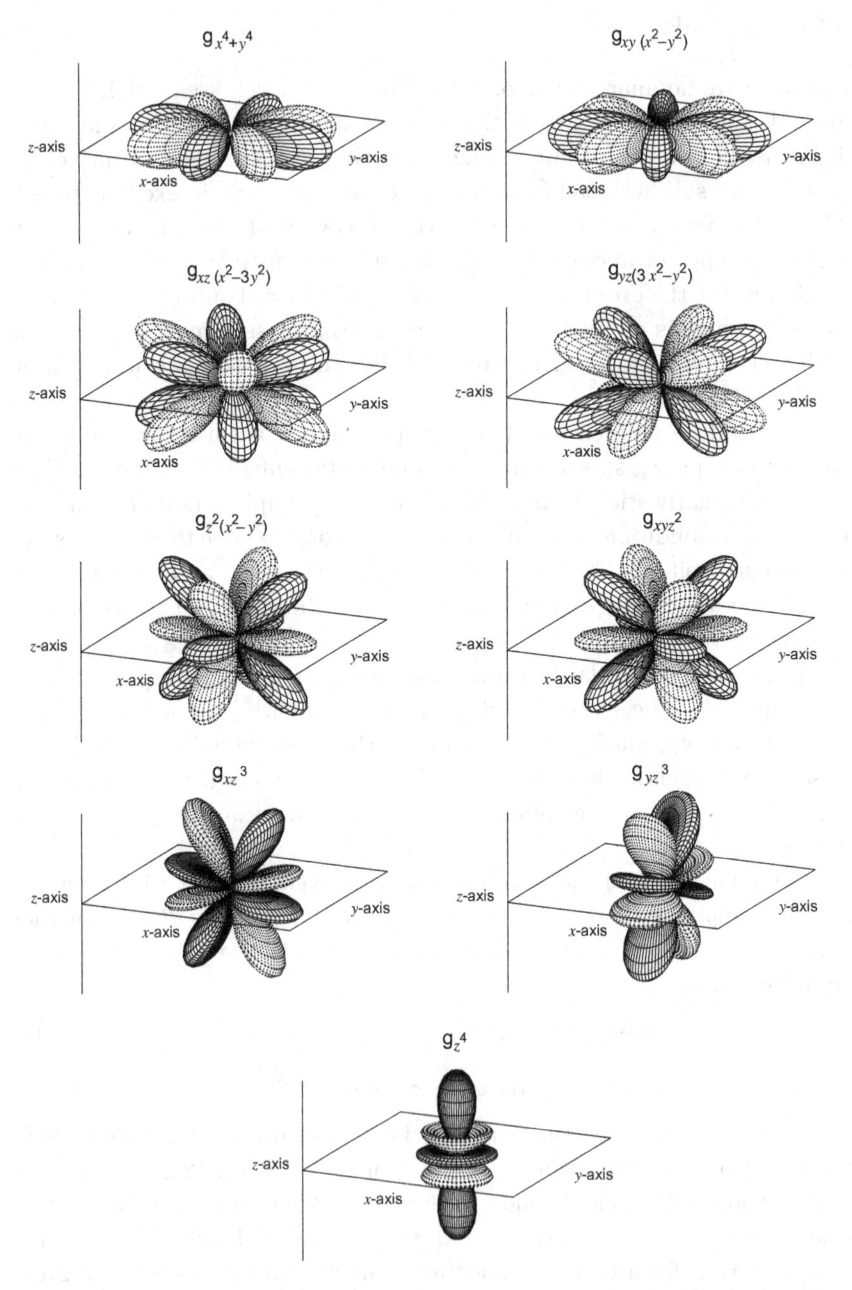

Figure 2.14. The angular wavefunctions for the atomic g orbitals. The broken lines indicate negative phase.

Table 2.9. The associated Legendre functions, P_4^m, with their normalisation constants, N_θ, relevant for g orbitals [25].

P_4^m	N_θ	$P_4^m(\theta)$
P_4^0	$9\sqrt{2}/16$	$((35/3)\cos^4\theta - 10\cos^2\theta + 1)$
$P_4^{\pm1}$	$9\sqrt{10}/8$	$\sin\theta((7/3)\cos^3\theta - \cos\theta)$
$P_4^{\pm2}$	$3\sqrt{5}/8$	$\sin^2\theta(7\cos^2\theta - 1)$
$P_4^{\pm3}$	$3\sqrt{70}/8$	$\sin^3\theta\cos\theta$
$P_4^{\pm4}$	$3\sqrt{35}/16$	$\sin^4\theta$

Table 2.10. The orbital angular wavefunctions in real form for g orbitals with their normalisation constants, N_φ, and the linear combinations of the $e^{im\varphi}$ functions required [25]. For the form of the $P_4^m(\theta)$ functions see Table 2.9.

g orbital	N_θ	$P_4^m(\theta)$	Linear comb.	N_φ	$\psi(\varphi)$
g_{z^4}	$9\sqrt{2}/16$	$P_4^0(\theta)$	$\|0\rangle$	$1/\sqrt{2\pi}$	1
g_{xz^3}	$9\sqrt{10}/8$	$P_4^1(\theta)$	$\|1\rangle + \|-1>$	$1/\sqrt{\pi}$	$\cos\varphi$
g_{yz^3}	$9\sqrt{10}/8$	$P_4^1(\theta)$	$\|1\rangle - \|-1>$	$1/\sqrt{\pi}$	$\sin\varphi$
$g_{z^2(x^2-y^2)}$	$3\sqrt{5}/8$	$P_4^2(\theta)$	$\|2\rangle + \|-2>$	$1/\sqrt{\pi}$	$\cos 2\varphi$
g_{xyz^2}	$3\sqrt{5}/8$	$P_4^2(\theta)$	$\|2\rangle - \|-2>$	$1/\sqrt{\pi}$	$\sin 2\varphi$
g_{x^3z}	$3\sqrt{70}/8$	$P_4^3(\theta)$	$\|3\rangle + \|-3>$	$1/\sqrt{\pi}$	$\cos 3\varphi$
g_{y^3z}	$3\sqrt{70}/8$	$P_4^3(\theta)$	$\|3\rangle - \|-3>$	$1/\sqrt{\pi}$	$\sin 3\varphi$
$g_{x^4+y^4}$	$3\sqrt{35}/16$	$P_4^4(\theta)$	$\|4\rangle + \|-4>$	$1/\sqrt{\pi}$	$\cos 4\varphi$
$g_{xy(x^2-y^2)}$	$3\sqrt{35}/16$	$P_4^4(\theta)$	$\|4\rangle - \|-4>$	$1/\sqrt{\pi}$	$\sin 4\varphi$

involve different m and so they are complex. The linear combinations of these four wavefunctions that produce real angular functions are given in Table 2.13.

The spherical and cubic orbitals are collected in Table 2.14. Three orbitals are common to both symmetries; these are indicated with an equals sign. The spherical labels denote the m labels used in linear combinations: $\pm m$ denotes the linear combination $|m\rangle + |-m\rangle$ and $\mp m$ denotes the linear combination $|m\rangle - |-m\rangle$. The octahedral label classifies the orbitals by how they are transformed by octahedral symmetry operations (two-, three- and four-fold rotations, etc.).

Table 2.11. The orbital angular functions for g orbitals in their full Cartesian form with normalisation constants. See the text for an explanation of the conversion.

g orbital	Normalisation	Full Cartesian function
g_{z^4}	$3/(16\sqrt{\pi})$	$(35z^4 - 30z^2r^2 + 3r^4)/r^4$
g_{xz^3}	$3\sqrt{10}/(8\sqrt{\pi})$	$xz(7z^2 - 3r^2)/r^4$
g_{yz^3}	$3\sqrt{10}/(8\sqrt{\pi})$	$yz(7z^2 - 3r^2)/r^4$
$g_{z^2(x^2-y^2)}$	$3\sqrt{5}/(8\sqrt{\pi})$	$(x^2 - y^2)(7z^2 - r^2)/r^4$
g_{xyz^2}	$3\sqrt{5}/(4\sqrt{\pi})$	$xy(7z^2 - r^2)/r^4$
g_{x^3z}	$3\sqrt{70}/(8\sqrt{\pi})$	$xz(x^2 - 3y^2)/r^4$
g_{y^3z}	$3\sqrt{70}/(8\sqrt{\pi})$	$yz(3x^2 - y^2)/r^4$
$g_{x^4+y^4}$	$3\sqrt{35}/(16\sqrt{\pi})$	$(x^4 + y^4 - 6x^2y^2)/r^4$
$g_{xy(x^2-y^2)}$	$3\sqrt{35}/(4\sqrt{\pi})$	$xy(x^2 - y^2)/r^4$

Table 2.12. The orbital angular wavefunctions for g orbitals in cubic symmetry in terms of spherical harmonic functions [28].

ψ(cubic)	$\psi(Y_4^m(\theta,\varphi))$
$\|\mathrm{g\ A_1\ a_1}\rangle$	$\frac{\sqrt{7}}{2\sqrt{3}}Y_4^0 + \frac{\sqrt{5}}{2\sqrt{6}}Y_4^4 + \frac{\sqrt{5}}{2\sqrt{6}}Y_4^{-4}$
$\|\mathrm{g\ E}\ \theta\rangle$	$-\frac{\sqrt{5}}{2\sqrt{3}}Y_4^0 + \frac{\sqrt{7}}{2\sqrt{6}}Y_4^4 + \frac{\sqrt{7}}{2\sqrt{6}}Y_4^{-4}$
$\|\mathrm{g\ E}\ \epsilon\rangle$	$\frac{1}{\sqrt{2}}Y_4^2 + \frac{1}{\sqrt{2}}Y_4^{-2}$
$\|\mathrm{g\ T_1\ 1}\rangle$	$-\frac{1}{2\sqrt{2}}Y_4^{-3} - \frac{\sqrt{7}}{2\sqrt{2}}Y_4^1$
$\|\mathrm{g\ T_1\ 0}\rangle$	$\frac{1}{\sqrt{2}}Y_4^4 - \frac{1}{\sqrt{2}}Y_4^{-4}$
$\|\mathrm{g\ T_1} - 1\rangle$	$\frac{1}{2\sqrt{2}}Y_4^3 + \frac{\sqrt{7}}{2\sqrt{2}}Y_4^{-1}$
$\|\mathrm{g\ T_2\ 1}\rangle$	$\frac{\sqrt{7}}{2\sqrt{2}}Y_4^3 - \frac{1}{2\sqrt{2}}Y_4^{-1}$
$\|\mathrm{g\ T_2\ 0}\rangle$	$\frac{1}{\sqrt{2}}Y_4^2 - \frac{1}{\sqrt{2}}Y_4^{-2}$
$\|\mathrm{g\ T_2} - 1\rangle$	$\frac{1}{2\sqrt{2}}Y_4^1 - \frac{\sqrt{7}}{2\sqrt{2}}Y_4^{-3}$

The Cartesian functions with normalisation constants for the cubic g orbitals are collected in Table 2.15. The cubic g orbital angular wavefunctions are plotted in Figure 2.15. Where the amplitude is negative the orbital is depicted with broken lines.

Table 2.13. Linear combinations of the cubic angular wavefunctions in the g subshell from Table 2.12 required to generate real functions, following the phases adopted by Griffith [28].

Cubic g orbital	Normalisation	Linear combination
$g_{xz(z^2-x^2)}$	$(1/\sqrt{2})$	$\lvert \mathrm{g\ T_1}\ 1\rangle + \lvert \mathrm{g\ T_1}\ -1\rangle$
$g_{yz(z^2-y^2)}$	$-\mathrm{i}(1/\sqrt{2})$	$\lvert \mathrm{g\ T_1}\ 1\rangle - \lvert \mathrm{g\ T_1}\ -1\rangle$
g_{xzy^2}	$(1/\sqrt{2})$	$\lvert \mathrm{g\ T_2}\ 1\rangle + \lvert \mathrm{g\ T_2}\ -1\rangle$
g_{yzx^2}	$\mathrm{i}(1/\sqrt{2})$	$\lvert \mathrm{g\ T_2}\ 1\rangle - \lvert \mathrm{g\ T_2}\ -1\rangle$

Table 2.14. The spherical and cubic g orbitals. See the text for the explanation of the labels.

Spherical g orbital	Label		Cubic g orbital	Label
g_{z^4}	0		$g_{x^4+y^4+z^4}$	A_1
g_{xz^3}	±1		$g_{xz(z^2-x^2)}$	T_1
g_{yz^3}	∓1		$g_{yz(z^2-y^2)}$	T_1
$g_{z^2(x^2-y^2)}$	±2	=	$g_{z^2(x^2-y^2)}$	E
g_{xyz^2}	∓2	=	g_{xyz^2}	T_2
g_{x^3z}	±3		g_{xzy^2}	T_2
g_{y^3z}	∓3		g_{yzx^2}	T_2
$g_{x^4+y^4}$	±4		$g_{2z^4-x^4-y^4}$	E
$g_{xy(x^2-y^2)}$	∓4	=	$g_{xy(x^2-y^2)}$	T_1

Table 2.15. The orbital angular functions for cubic g orbitals in their full Cartesian form with normalisation constants.

g orbital	Normalisation	Full Cartesian function
$g_{x^4+y^4+z^4}$	$\sqrt{21}/(4\sqrt{\pi})$	$(x^4+y^4+z^4-3(x^2y^2+x^2z^2+y^2z^2))/r^4$
$g_{2z^4-x^4-y^4}$	$\sqrt{15}/(8\sqrt{\pi})$	$(2z^4-x^4-y^4+6(2x^2y^2-x^2z^2-y^2z^2))/r^4$
$g_{z^2(x^2-y^2)}$	$3\sqrt{5}/(8\sqrt{\pi})$	$(x^2-y^2)(7z^2-r^2)/r^4$
$g_{xz(z^2-x^2)}$	$3\sqrt{35}/(4\sqrt{\pi})$	$xz(z^2-x^2)/r^4$
$g_{yz(z^2-y^2)}$	$3\sqrt{35}/(4\sqrt{\pi})$	$yz(z^2-y^2)/r^4$
$g_{xy(x^2-y^2)}$	$3\sqrt{35}/(4\sqrt{\pi})$	$xy(x^2-y^2)/r^4$
g_{xyz^2}	$3\sqrt{5}/(4\sqrt{\pi})$	$xy(7z^2-r^2)/r^4$
g_{xzy^2}	$3\sqrt{5}/(4\sqrt{\pi})$	$xz(7y^2-r^2)/r^4$
g_{yzx^2}	$3\sqrt{5}/(4\sqrt{\pi})$	$yz(7x^2-r^2)/r^4$

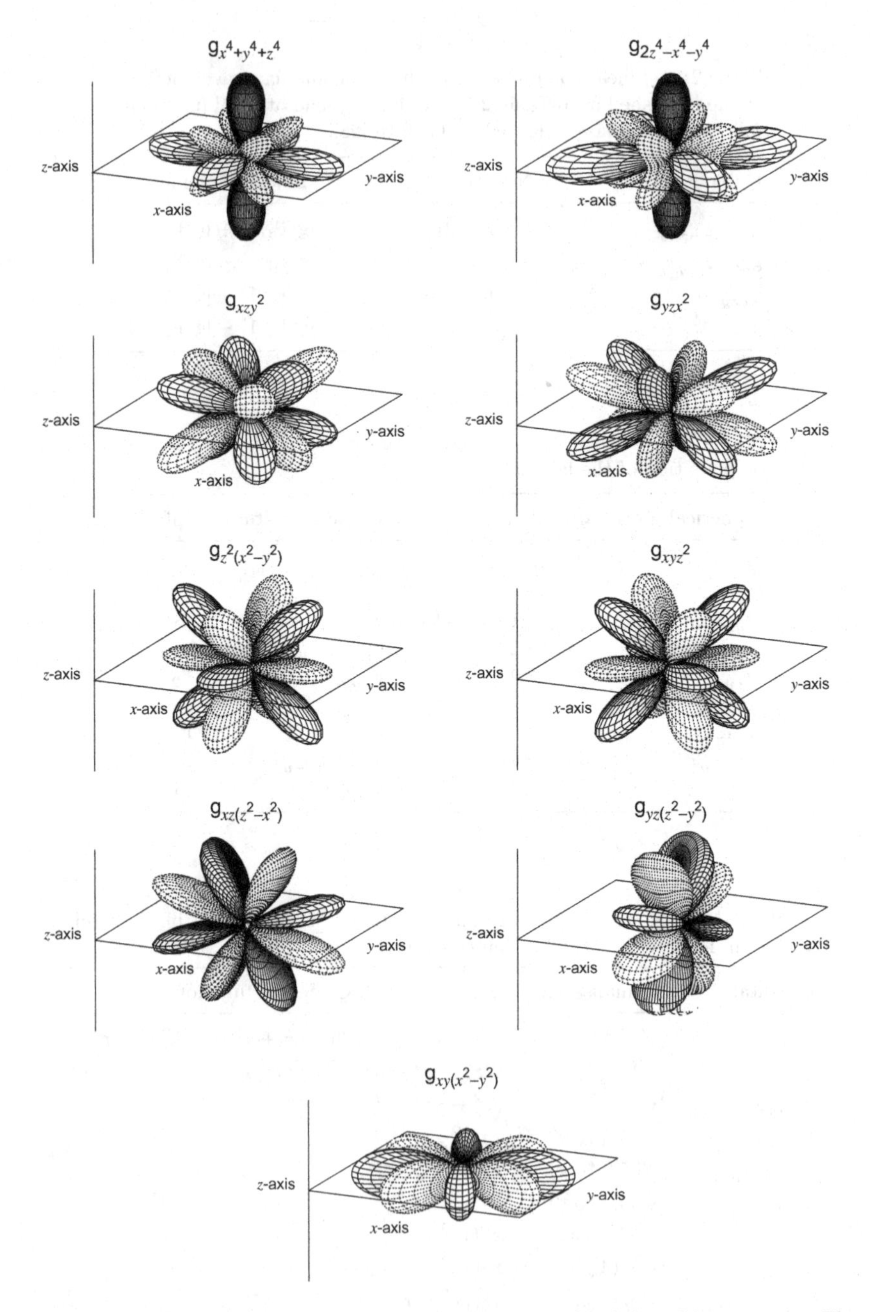

Figure 2.15. The angular wavefunctions for the atomic g orbitals in cubic symmetry. The broken lines indicate negative phase.

2.10 The Speed of the Electron

If we divide the momentum operator, $\hat{p} = \frac{\hbar}{i}\nabla$, by the electron mass then we have a velocity operator. The hydrogen orbital wavefunctions are not, generally, eigenfunctions of the velocity operator though, so the electrons do not have a well-defined velocity. However, an expectation value of electron velocity may be calculated using equation (2.39). The first derivative changes the parity of the wavefunction, so the integral $\langle \psi \mid v \mid \psi \rangle$ must be zero. This isn't surprising since velocity is a vector so positive and negative values would be expected to cancel out.

Exercise 32

Show that the expectation value for the velocity of a 1s electron is zero.

An alternative method to find the magnitude of electron velocity is to deduce it from the electron's kinetic energy, T, since $v = \sqrt{2T/\mu}$, where μ is the reduced mass of the electron. Being a scalar, the kinetic energy should be easier to find — especially for s electrons as there is no rotational kinetic energy to calculate. Furthermore, the hydrogen orbital wavefunctions are eigenfunctions of the kinetic energy operator, $\hat{T} = -\frac{\hbar^2}{2\mu}\nabla^2$.

Exercise 33

Use the virial theorem (equation (2.41)) to show that the kinetic energy of a hydrogen 1s electron is found using minus the expression in equation (2.33). Show that this energy is consistent with the experimentally determined ionisation energy for hydrogen of 1312 kJ mol^{-1} [31].

Exercise 34

Calculate the speed of a hydrogen electron in a 1s orbital using the observed ionisation energy of 1312 kJ mol^{-1} and the virial theorem (equation (2.41)). Express the speed as a percentage of the speed of light.

Atoms with only one electron but more than one proton (i.e. ions), such as He^{+}, Li^{2+}, Be^{3+}, etc., are known as hydrogenic as they can be understood using the Schrödinger equation for hydrogen, except with a higher value of Z. Equations (2.33) and (2.41) together predict that v^2 should increase in proportion with Z^2, hence electron speed should increase linearly with atomic number. The ionisation energies of the hydrogenic atoms are recorded in Moore's tables [31, 41] up to $Z = 29$, Cu^{28+}.

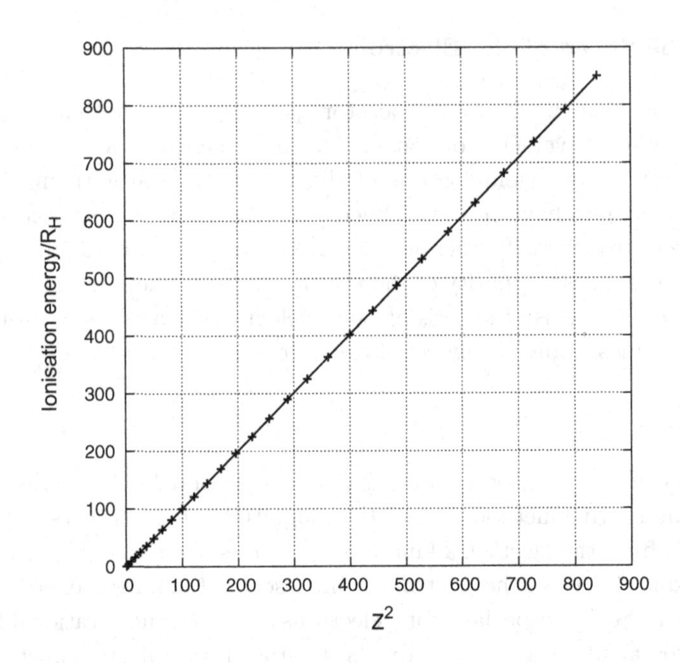

Figure 2.16. Observed hydrogenic atom ionisation energies plotted against Z^2 (data from [31, 41]).

The ionisation energies, measured in units of the Rydberg energy, are plotted against Z^2 in Figure 2.16.

It can be seen that there is little deviation from linearity. Equating $\frac{1}{2}\mu v^2$ to $\frac{Z^2 e^2 K}{2n^2 a_0}$, a speed for the electron can be calculated for hydrogenic atoms. Since the speed of the electron in hydrogen is 0.730% of the speed of light, the value for Z at which the electron speed is predicted to equal the speed of light is $Z = 100/0.73 = 137$. This of course is the reciprocal of the fine structure constant, $\alpha = e^2 K/\hbar c$, the dimensionless electromagnetic constant. Expressing the kinetic energy of the 1s electron of a hydrogenic atom in terms of α using equations (2.26), (2.33) and (2.41) gives the following:

$$E_{\text{kin}} = \frac{1}{2}\mu v^2 = \frac{1}{2}\mu c^2 Z^2 \alpha^2. \tag{2.45}$$

It follows that $v = cZ\alpha$. If an electron were moving at the speed c, then Z would have to be 137, the reciprocal of the fine structure constant.

Equation (2.45) above is non-relativistic, i.e. it does not take into account Einstein's special theory of relativity. When speeds become a significant proportion of the speed of light, Einstein's theory demands that a mass correction needs to be made, by applying his gamma term:

$$\gamma = \frac{1}{\sqrt{1 - \frac{v^2}{c^2}}}. \tag{2.46}$$

Substituting $v = cZ\alpha$ into the expression for γ and multiplying the electron rest reduced mass, μ_0, by γ, produces the following corrected relativistic expression for the kinetic energy of the electron:

$$E_{\mathrm{kin,rel}} = \frac{\frac{1}{2}\mu_0 c^2 Z^2 \alpha^2}{\sqrt{1 - Z^2\alpha^2}}. \tag{2.47}$$

For $Z > 137$, $E_{\mathrm{kin,rel}}$ is imaginary. For this reason, Feynman suggested that the periodic table must end here, regardless of nuclear stability (the usual limiting factor). It has been proposed that relaxing the assumption that the nucleus is a point charge gets around this impasse. The issue has been discussed by Philip Ball [32]. While the preceding argument is based on a one-electron atom, it will still serve as an approximation for many-electron atoms since electrons outside the 1s subshell provide very little shielding from the nucleus for a 1s electron (see Section 3.3).

It should be emphasised that the application of the relativistic correction to the electron mass does not make this approach formally relativistic. To incorporate special relativity rigorously the equations need to be Lorentz-invariant: i.e. the Dirac equation needs to be solved, which is beyond the scope of this book. However, a relativistic mass correction gives an approximation of the relativistic effect within the Schrödinger framework. This effect is important for the heaviest elements, especially in the d block, and will be considered further in Section 3.10.

Chapter 3

Multi-Electron Atoms

3.1 Multi-Electron Schrödinger Equations and the Orbital Approximation

If we take the same approach for helium that we did for the hydrogen electron with the Schrödinger equation, then we need to adjust Z to 2, take into account the extra mass of the nucleus in μ, but most importantly new terms appears in the operator. While the Schrödinger equation for the hydrogen electron has essentially two terms, one for the kinetic energy and one for the electron–nuclear potential energy (see equation (2.22)), an extra three terms appear in the helium operator: the kinetic and electron–nuclear potential energy terms for the second electron and a repulsive electron–electron potential energy term, $\hat{V}_{12}$, with the form

$$\hat{V}_{12} = \frac{e^2 K}{\sqrt{(x_1 - x_2)^2 + (y_1 - y_2)^2 + (z_1 - z_2)^2}}, \tag{3.1}$$

where the denominator is the distance between the two electrons and the subscripts 1 and 2 refer to electrons 1 and 2. Unfortunately analytical eigenfunctions (such as associated Laguerre functions, etc.) are not known for this Schrödinger equation, nor for ones with more electrons. As a result we make the 'orbital approximation' that we may use one-electron wavefunctions as the basis for the wavefunctions of multi-electron atoms.

3.2 Shielding and Penetration

The effect of having extra electrons, due to electrostatic repulsion, is to shield a given electron to some extent from the nucleus. The extent to which one electron shields another can be estimated from the radial density functions of the two electrons. If one electron is much closer to the nucleus than another, then the outer electron will feel the charge of the nucleus diminished by nearly a unit of charge due to the shielding of the inner electron. The inner electron, by contrast, will experience very little shielding from the outer electron. Generally electrons are well shielded by others that are part of inner principal shells, due to the smaller radial extent of electrons from inner shells (see Figure 3.1).

The radial density functions for the three subshells in the third shell of the hydrogen atom are shown in Figure 3.1. Within a given shell in the hydrogen atom, there is a lower $\langle r \rangle$ associated with subshells of higher l (see equation (2.42)). From this one might expect subshells with higher l preferentially to shield those of lower l in multi-electron atoms, leaving the higher-l subshells at lower potential energy. In fact, the opposite ordering of subshell energies in multi-electron atoms is observed:

$$\mathrm{s} < \mathrm{p} < \mathrm{d} < \mathrm{f} \ldots$$

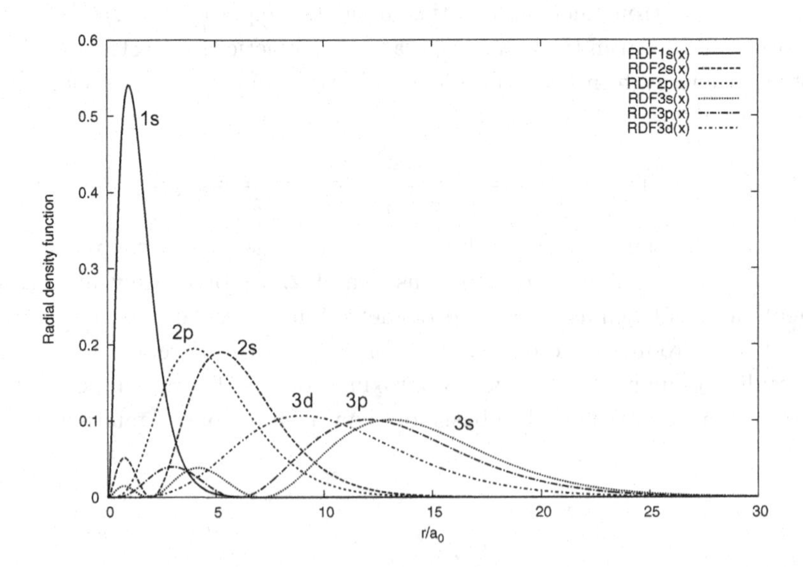

Figure 3.1. Radial density functions for all the H atom subshells of the first three shells.

This is because of an effect known as penetration. Orbitals with a lower value of l have more radial nodes, $n - l - 1$, which leads to more electron density very close to the nucleus. In this region the electrons are not shielded effectively by inner-shell electrons (see Figure 3.1), increasing the nuclear charge experienced and lowering the potential energy. The apparent order of the valence subshells (with some exceptions) in terms of ground-state electron configurations in free, i.e. gas-phase, atoms as atomic number increases, known as the aufbau principle, is as follows:

$$1\text{s} \ll 2\text{s} < 2\text{p} \ll 3\text{s} < 3\text{p} \ll 4\text{s} < 3\text{d} < 4\text{p} \ll 5\text{s} < 4\text{d} < 5\text{p} \ll 6\text{s} \ldots$$

The energy gaps are particularly large after the p subshells, which is why they are considered to close the valence shells.

The ordering of all the subshells implies that the degeneracy of subshells within a principal shell that was seen in the hydrogen atom no longer applies to multi-electron atoms. However, the degeneracy of orbitals with a subshell still applies for multi-electron atoms (while they are free atoms in the gas phase).

This ordering only relates to the sequence of ground-state gas-phase electron configurations as Z increases in the periodic table. It is frequently incorrectly interpreted as providing the ordering of subshell energies in all atoms and ions. Alternative orderings of the subshells are considered at the end of Section 3.9.

This ordering follows the Madelung rule that subshells are ordered by their increasing $n + l$ value; where subshells have the same $n + l$ value they are ordered by their increasing n value. A theoretical justification for this ordering has been proposed by Bent and Weinhold [33]. They equate the number of nodes to $n + l$, where they include $r = \infty$ as a radial node and define the number of angular nodes as the number of sign changes in a revolution, which equals $2l$ rather than l. So rather than the conventional total of nodes of $n - l - 1 + l = n - 1$, this alternative method gives $n - l - 1 + 1 + 2l = n + l$, the Madelung rule.

The s subshells have the greatest penetration and so the reduced shielding they experience accounts for their early positions in the aufbau series for free atoms. This series of subshells accounts for the shape of the periodic table. It is remarkable that the 4p^1 excited state and even the 5s^1 excited state of potassium are lower in energy than the 3d^1 excited state [31], showing that even the 4p and 5s orbitals have sufficient penetration inside the 3d orbital for them to have lower energy than the 3d orbital when they are occupied.

An elegant model was adopted by Sommerfeld to account for the penetration of s electrons, namely that of elliptical orbits. The electron orbits are considered classically, analogously to planetary orbits, before the constant angular momentum is quantised. This model generates two quantum numbers: a radial one, n_r, and an azimuthal one, k [25]. The distance of closest approach, r_{min}, of the electron to the nucleus can then be found in terms of the total quantum number, n, which is defined as the sum of n_r and $|k|$:

$$r_{\text{min}} = \frac{n(n - \sqrt{n^2 - k^2})}{Z} a_0, \tag{3.2}$$

where a_0 is the Bohr radius and Z is the proton number. The equation predicts, not surprisingly, that with increasing Z there is a decrease in r_{min}. The equation shows that r_{min} takes its smallest value, corresponding to maximum penetration, when the magnitude of k, i.e. angular momentum, is at a minimum. The orbit with maximum penetration will be the most eccentric orbit, and so also have the greatest r_{max}. By contrast, the orbit with maximum angular momentum will have zero eccentricity and be circular, i.e. with a constant r that is intermediate between the r_{min} and r_{max} of eccentric, lower angular momentum orbits. This model reproduces the qualitative behaviour of different shells and subshells.

It is easy to get oneself into a tangle when considering the relative energies of orbitals in multi-electron atoms: there has, for example, been much argument about the relative energies of 4s and 3d orbitals in first-row d block metals [34] and this question will be considered in Section 3.7. Scerri has made the point that arguments and calculations based on orbital energies are, in any case, relying on the orbital approximation [35]. It must be remembered that orbital energies depend on the electron configuration as well as the proton number [36].

3.3 Slater's Rules

In the Slater approach one can estimate the energy of an electron in an orbital based on the effective nuclear charge, Z_{eff}, experienced by the electron, which takes into account the electron's interactions with other electrons in the atom. Strictly, however, as is discussed in Section 4.1, we have to consider the wavefunction for the atom as a whole: the electronic wavefunction is not just a simple sum or product of the wavefunctions of the individual electrons. This leaves us in difficulty when trying to describe the energies of single electrons in a multi-electron atom. One meets further problems when trying to rationalise the energy of an orbital in an atom that isn't occupied.

In 1930 Slater devised some rules semi-empirically to quantify the Z_{eff} experienced by an electron in a multi-electron atom. The energy of a lone electron in the valence shell can be approximated to the energy required to ionise the electron, i.e. to take it from its shell, n, to $n = \infty$ (this is Koopmans' theorem). The Z_{eff} of a lone valence electron can therefore be approximated experimentally from solving equation (2.34) for Z, with E_n being minus the ionisation energy of the electron in question. Z_{eff} is related to Z through the shielding constant, S:

$$Z_{\text{eff}} = Z - S. \tag{3.3}$$

The situation is less straightforward when there are multiple electrons in the valence shell. This is because the ionisation of one of the valence electrons changes the shielding experienced by the other valence electron(s). By contrast, the energy of the core electrons is affected little by the ionisation of a valence electron. In order to estimate the ionisation energy of an atom with multiple valence electrons, the total energy of the valence electrons of the atom before ionisation must be subtracted from the total energy of the valence electrons after ionisation (we are assuming the core electron energies are unaffected).

With reference to ionisation energy data, Slater was able to formulate rules to give an approximate shielding constant, and therefore Z_{eff} [37]. Every other electron in the atom needs to be considered in turn when working out the shielding contribution, σ, for each other electron; the shielding factors from all the other electrons are summed to give S: $S = \Sigma\sigma$. The rules devised by Slater for finding the shielding contribution from each of the other electrons to the one under consideration are shown in the box [38]. We see in his first rule that Slater did not follow the Madelung rule; his ordering of subshells is how they are radially arranged in multi-electron atoms.

Slater's Rules

(1) Write the electron configuration according to the following groupings (in parentheses) in the following order:
(1s)(2s 2p)(3s 3p)(3d)(4s 4p)(4d)(4f)(5s 5p)(5d)(5f)...
(2) $\sigma = 0$ for electrons in higher groupings.
(3) $\sigma = 0.35$ for electrons in the same grouping as the electron under consideration, except for 1s when $\sigma = 0.30$.
(4) For s and p electrons: $\sigma = 0.85$ for electrons with n one less, $\sigma = 1.00$ for electrons with n two or more less.
(5) For d and f electrons: $\sigma = 1.00$ for all electrons in lower groupings, reflecting their low penetration.
(6) $Z_{\text{eff}} = Z - \Sigma\sigma$.
(7) In the fourth, fifth and sixth shells, $n_{\text{eff}} = 3.7$, 4.0 and 4.2, respectively, which are required for energy calculations.

Due to the imperfect shielding of valence electrons, the energy of any occupied orbital will decrease as the atomic number, Z, increases. Indeed d and f electrons are so inefficient at shielding s and p electrons in the same shell that they make no contribution to their shielding in Slater's rules. This can be rationalised in terms of radial nodes. The d and f radial wavefunctions have fewer nodes than s and p wavefunctions in the same shell (the 3d and 4f radial wavefunctions have no nodes at all). This means that there is significantly less d and f electron density close to the nucleus than is found for s and p electrons in the same shell, reducing their shielding of these subshells.

The Rydberg equation as seen in equation (2.34) can be used with Slater's rules to estimate the energy of electrons in multi-electron atoms by using Z_{eff} in place of Z.[1] In the case of hydrogen and other one-electron atoms, following ionisation no electron remains in the atom and so the ionisation energy is just equal to minus the energy of the electron. Similarly, in atoms with a single electron in one of Slater's bracketed grouping of orbitals (see his rules in the box above), the ionisation energy calculation

[1]The other adaptation of the Rydberg equation, using the quantum defect, is described in Section 6.3.

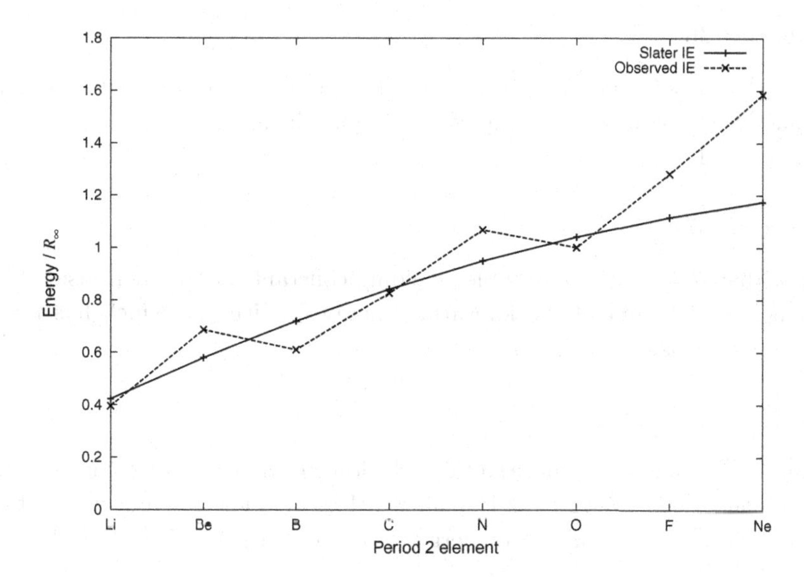

Figure 3.2. Slater and observed [39] ionisation energies in Rydbergs, R_∞, for period 2 atoms.

is very simple because the outer electron (assuming that is the one being ionised) does not affect the energies of the inner electrons as it is not judged by Slater's rules to shield them at all.

The ionisation energies predicted by Slater's rules for the period 2 atoms are plotted in Figure 3.2 along with the observed ionisation energies [39], measured in Rydbergs, R_∞, i.e. referenced to 1 for the ionisation energy of a hydrogen atom. R_∞ is the Rydberg constant using the rest mass of the electron rather than the reduced mass, since it is being used for elements heavier than hydrogen.

The agreement between the Slater ionisation energies and the observed ones in Figure 3.2 is surprisingly good considering that Slater's rules make no distinction between s and p electrons in the same shell and take no account of spin-pairing (see Section 3.8). Agreement is much less good for subsequent periods, however, especially longer ones.

Exercise 35

Using the Rydberg formula of equation (2.34) and Z_{eff} from Slater's rules, calculate the ionisation energy of oxygen.

Exercise 36

Show that $\sigma = 0.30$ is the optimal shielding constant within the 1s subshell to model the ionisation energy for helium, which has been measured at 2372 kJ mol^{-1} [39].

Exercise 37

Show that $\sigma = 0.85$ is a suitable shielding constant for the shell inside the valence shell to model the ionisation energy for lithium, which has been measured at 520 kJ mol^{-1} [39].

Exercise 38

Show that $\sigma = 0.35$ is a more suitable shielding constant for electrons within the second valence shell than is 0.30 (which is used for electrons within the first shell). Use the ionisation energy for oxygen (1314 kJ mol^{-1} [39]).

Exercise 39

Work out the Z_{eff} for the outer electron and the ionisation energy for all the alkali metals using Slater's rules and equation (2.34). Compare your answers to the experimentally determined values (in kJ mol^{-1}) [39, 40]: Li 520, Na 496, K 419, Rb 403, Cs 376, Fr 380.

A common error is to attribute the decrease in ionisation energy on descending group 1 to a decrease in Z_{eff}. In fact, on descending from the third period the Slater Z_{eff} remains constant on account of the additional core electrons two shells beneath the outer shell balancing the additional protons in the nucleus. The dominating factor in the reduction of observed ionisation energies on descending a group is the n_{eff}^2 on the denominator of the Rydberg equation.

Slater's rules are useful for establishing general trends rather than the exact values or subtleties in trends [38]. For example, there is nothing in Slater's rules to anticipate the drop in ionisation energies between the s and p blocks in periods 2 and 3, nor between the 15th and 16th groups due to the pairing of electrons in the p subshell.

3.4 Electron Affinity and its Periodicity

While ionisation energy data have been known since the 1950s [31, 41, 42], prior to 1970 only the electron affinities of hydrogen, fluorine, chlorine, bromine, iodine, carbon, oxygen and sulfur had been definitively measured [43]. For this historical reason, electron affinities are often given less emphasis in discussions of atomic structure than ionisation energies. Modern values for the atomic electron affinities are collected in Table 3.1. A well-known website [40] consulted by the author agreed on the electron affinities, except for some revisions given in a more modern source [44].

There has also been some argument about the sign convention, which general textbooks conventionally give as an electron attachment energy. Wheeler has argued for a return to the original convention that defined the energy as the electron detachment energy, since it is also followed in the more recent experimental and theoretical literature [45]. Such a change would also be consistent with the word 'affinity' and the convention for ionisation energy, which would make the electron affinity effectively a zeroth ionisation energy. This would make sense given the similar periodicity displayed by electron affinity and ionisation energies (discussed later in this section). This book will therefore follow the electron detachment energy convention.

A striking feature of the electron affinities in Table 3.1 is that elements whose subshells are closed, i.e. those in groups 2, 12 and 18, usually take a value of zero. This is on account of the effective screening of closed subshells. Nitrogen and, to a lesser extent, the other group 15 elements have low electron affinities, which may be due to the exchange stabilisation of the half-full subshell (discussed in Section 4.2). A similar effect is seen in the transition metals in group 7. Palladium's low electron affinity compared with other members of group 10 is consistent with its having the only d^{10} configuration in that group. The zero for hafnium is more mysterious; Myers has suggested it may be an error of measurement [46].

Electron affinity shows a similar periodicity to successive ionisation energies. This can be made particularly clear by using the following measure of effective nuclear charge, Z_{eff}, based on the Rydberg equation (2.34) (similar to the approach taken by Wheeler [45]).

$$\text{IE} = \text{IE(H)} \times \frac{Z_{\text{eff}}^2}{n_{\text{eff}}^2} \tag{3.4}$$

$$Z_{\text{eff}} = n_{\text{eff}} \sqrt{\frac{\text{IE}}{\text{IE(H)}}}. \tag{3.5}$$

Table 3.1. Electron affinities for the atomic elements in kJ mol^{-1}, expressed as detachment energies [40, 44].

Element	EA	Element	EA	Element	EA
H	73	Mn	0	In	39
He	0	Fe	15	Sn	107
Li	60	Co	64	Sb	101
Be	0	Ni	112	Te	190
B	27	Cu	119	I	295
C	122	Zn	0	Xe	0
N	−7	Ga	39	Cs	46
O	141	Ge	119	Ba	14
F	328	As	79	La	45
Ne	0	Se	195	Hf	0
Na	53	Br	325	Ta	31
Mg	0	Kr	0	W	79
Al	42	Rb	47	Re	14
Si	134	Sr	5	Os	104
P	72	Y	30	Ir	151
S	200	Zr	41	Pt	205
Cl	349	Nb	86	Au	223
Ar	0	Mo	72	Hg	0
K	48	Tc	53	Tl	36
Ca	2	Ru	101	Pb	35
Sc	18	Rh	110	Bi	91
Ti	8	Pd	54	Po	183
V	51	Ag	126	At	270
Cr	65	Cd	0	Rn	0

Z_{eff} is deduced from the comparison between the ionisation energy of the atom or ion of interest and hydrogen. n_{eff} is the effective shell number following Slater's rules.

The advantage of this approach is that the large increases of ionisation (or electron detachment) energy on ionisation of atoms are manifest as only small increases in effective nuclear charge, allowing for easy visualisation in a single figure. The effective nuclear charges are plotted in Figure 3.3.

It is clear from Figure 3.3 that the electron affinity data share the same trend as the familiar one seen in the atoms and 1+ ions, except displaced such that isoelectronic species are comparable. This justifies the Mulliken definition of electronegativity based on the sum IE + EA as, when combined, the irregularities in the trends, occurring at different elements, are smoothed out. For the one-electron species He^+, $Z_{\text{eff}} = Z$ as there is no shielding. The scheme is less useful for the transition metals due to the irregularities

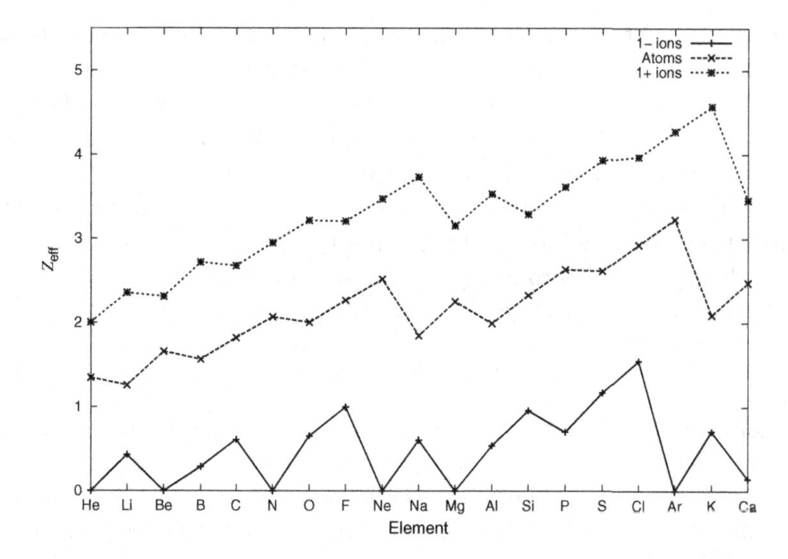

Figure 3.3. Effective nuclear charges for atoms, 1+ and 1− ions from equation (3.5) (using electron affinity and ionisation data from [39, 40, 44]).

in the electron configurations of their atoms and ions and the consequent ambiguity in the appropriate value of n to use.

Exercise 40

Calculate the Mulliken electronegativity, χ, of oxygen using Slater's rules and also using the observed ionisation energy (1314 kJ mol^{-1} [39]) and electron affinity (141 kJ mol^{-1} [40, 44]). Use the following expression with the ionisation energy (IE) and electron affinity (EA) in kJ mol^{-1}:

$$\chi = 1.97 \times 10^{-3}(\mathrm{IE} + \mathrm{EA}) + 0.19. \tag{3.6}$$

3.5 Electron Spin and the Pauli Principle

Until now no mention has been made of electron spin, as it has little effect on the atomic spectrum of hydrogen, but it has far-reaching impact with multi-electron atoms. Electrons, like protons and neutrons, are fermions, which means they behave as if they carry half a unit of intrinsic angular momentum, i.e. $s = \frac{1}{2}$ in $\hbar\sqrt{s(s+1)}$. The words 'behave as if' are used because electrons are believed to be quantum ('smeared') points, which gives them no real scope to produce an angular momentum by rotating about their own axis of symmetry as a classical body would. (If they were rotating around their own axis their spin quantum number would be an integer.) There is no classical analogue of electron spin.

Electron spin was discovered experimentally in the Stern–Gerlach experiment of 1922, explained in [47], and is also predicted by Dirac's relativistic theory. Rowlands has pointed out that, while not predicted by the non-relativistic Schrödinger equation in its traditional form, spin can emerge from this equation if the full product of the ∇ operator is used rather than the scalar product [48]. The full product of two vectors **a** and **b**, which is not commutative, is:

$$\mathbf{ab} = \mathbf{a} \cdot \mathbf{b} + \mathrm{i}\mathbf{a} \times \mathbf{b}. \tag{3.7}$$

The combination of the electron spin with its charge results in a magnetic moment. There is an additional quantum number, m_s, that gives the projection of the spin angular momentum on the z-axis in units of $\hbar$. It may take the values $+\frac{1}{2}$ or $-\frac{1}{2}$. Apart from being half-integral, m_s is analogous to the quantum number m describing the projection of the orbital angular momentum l. This will be called m_l from now on, distinguishing it from m_s. The properties of electron spin are considered in detail in [49]. The quantum theory of angular momentum discussed in Section 2.3 also applies to electron spin.

The Pauli principle postulates that no two electrons in an atom may have all their quantum numbers identical.[2] The quantum numbers n, l and m_l effectively identify the orbital (allowing for linear combinations of m_l). Since all electrons have the same value of s and can only adopt one of two possible m_s values, then the Pauli principle determines that the maximum electron-occupancy of an orbital is two, and only when the electrons have different values of m_s. The plus and minus half values of m_s are often referred to as 'spin-up' and 'spin-down', respectively; their wavefunctions are often abbreviated as α and β, respectively.

While the spin vectors may point up and down on average, the instantaneous angle of the spin vector to the vertical, θ, is found from $\cos\theta = \frac{m_s}{\sqrt{s(s+1)}}$. This gives $\theta = 54.7°$ (the magic angle in NMR), so the 'up' and 'down' spin vectors are mainly horizontal. Specifying the s_z component of the spin vector means that s_x and s_y cannot be known exactly. However, an equatorial s_{xy} vector must, by Pythagoras, have magnitude $\hbar\sqrt{s(s+1) - m_s^2}$.

[2] Any one-particle state can be occupied by no more than one fermion but many bosons.

It has been (flippantly) said that the Pauli principle accounts for all of chemistry, since without it all electrons would be 1s electrons. Together with the ordering of the subshell energies given earlier and their sequential filling according to the Aufbau principle, the Pauli principle gives the electron configuration of an atom. The origin of the Pauli principle is in the antisymmetry of multi-electron wavefunctions upon exchange of any two electrons, discussed in Section 4.1.

Exercise 41

What is the total number of electrons that can occupy a principal quantum shell, n?

3.6 Arrow-in-Box Notation and its Limitations

A typical way to approach multi-electron configurations is through arrow-in-box notation, commonly taught to pre-university students: each box represents an orbital that can accommodate up to two electrons, as long as they are spinning in opposite senses, represented with up- and down-pointing arrows (to satisfy the Pauli principle). This notation is usually met after considering the order of subshell energies in multi-electron atoms. Successive electrons are shown to occupy different orbitals, where possible, in the process of filling a subshell to arrive at an electron configuration — based on the reasonable assumption that occupying different orbitals reduces the repulsive interactions between electrons. In addition to this, successive electrons are shown, where possible, to be spinning in parallel (conventionally spin-up for the first electrons in a subshell). These parallel spins may just be taken to be a convention at this level; some students may, however, learn that there is an energetic advantage to having the electrons spinning in parallel. This is the so-called exchange energy, which is explained in detail in Section 4.2.

An advantage of arrow-in-box notation is that it provides an intuitive visual explanation for the exceptional drop in ionisation energy observed when passing from p^3 to p^4 configurations: as p^4 is the first p^n configuration with a paired electron, it is the one experiencing the additional repulsion from orbital-sharing electrons (which the previous element doesn't).

The problem with the arrow-in-box picture is that it promotes the idea that electrons can be treated independently in multi-electron atoms — an assumption that can be particularly problematic in transition metals, as we shall see. Atomic spectra can only be rationalised in terms of wavefunctions that represent states where the electrons in an atom have coupled to form a many-electron resultant. Furthermore, the wavefunctions have to be antisymmetrised, as explained in Section 4.1. i.e. A multi-electron wavefunction is not simply the product of the wavefunctions for individual electrons, which is implied by the arrow-in-box notation.

The ground-state of the d^5 configuration might be taken to be simply five spin-up electrons, one in each d orbital, with perhaps the five electrons spin-down as a second possible microstate. As is explained in Sections 4.1 and 4.3, n spin-parallel electrons in a subshell will be $(n+1)$-fold degenerate, so there are actually six possible microstates for the ground state of d^5, each of which will be a (5×5) determinant, expanding out to $5! = 120$ terms.

3.7 Ionisation of d-Block-Metal s Electrons

A peculiarity of the d block metals is that the valence s and d electrons ionise from the gas-phase atoms in the reverse order of what might be predicted by the aufbau principle. This is the case for all three rows of the d block. We saw in Section 3.2 that the final electrons to enter the 19th and 20th elements in the periodic table enter the 4s subshell rather than the 3d. This is due to the extra penetration of the 4s orbital reducing the shielding it experiences from inner electrons and reducing its energy compared with the 3d subshell. At the start of the 3d series, however, the 3d subshell energy drops below the 4s energy [34, 50], leading to the ionisation of electrons from the 4s subshell, as observed. This is because the 4s electrons do not effectively shield the 3d orbitals, since the latter have a significantly lower average radius, causing the 3d energy to drop rapidly. The same effects are seen in the second and third rows of the d block. This raises the question as to why the electron configurations of d block metal gas-phase atoms don't change to reflect the lower energy of the 3d electrons. The 3d subshell is more compact than the 4s subshell and so is more sensitive to repulsive electron–electron interactions which raise their electrons' energies. This can be alleviated by promoting electrons to the 4s subshell where they are more distant [51].

The preceding discussion brings into focus two common misconceptions relating to the aufbau principle. The first relates to the belief that individual electrons in an atom can be considered independently (see Section 3.6). Subshell energies, in the context of multi-electron atoms, are specific to the electron configuration as well as the proton number: they cannot be assumed to remain the same — even relative to one another — if an electron is moved from one subshell to another. This explains the apparent paradox of 3d being lower in energy than 4s for first-row d block metals which have 4s electrons in their ground state in the gas phase despite an incomplete 3d subshell. It should be emphasised that s orbital occupancy is lower in the more chemically relevant condensed phases of metals, and in transition metal complexes.

The second misconception is that the aufbau ordering applies to ions as well as neutral atoms, rather than just being a tool to predict the ground-state electron configuration of gas-phase atoms. It is tempting to talk in terms of 'filling the atom' with electrons, as if one had a bare nucleus and added to it one electron at a time, but the aufbau ordering of subshells applies only to neutral atoms, and even then only as a predictor of electron configuration, rather the relative energy of the subshells, as shown in the last paragraph. In the case of titanium ions, its 3+ ion has the configuration [Ar] $3d^1$, Ti^{2+} is $3d^2$ and Ti^+ is $3d^2\ 4s^1$. This shows that after the electrons have filled the orbitals up to [Ar] in a Ti^{4+} ion it is the 3d subshell that is next occupied, not 4s as might be predicted from the aufbau ordering. However, the sequence of adding all the d electrons followed by the s electrons to the noble gas core in titanium is not perfectly replicated by all the other first-row d block metal ions, as shown by the data in Table 3.2. Neither are these configurations consistent with all those in the second and third rows of the d block, as is made clear in Tables 3.3 and 3.4.

Despite the variation between the three series of d block elements it is clear from Tables 3.2, 3.3 and 3.4 that the d subshell drops in energy more quickly than the s subshell as the atom becomes more ionised: there are no known cases of 2+ d block metal ions with its valence s subshell occupied in its ground state, even in group 12 (Zn, Cd, Hg) where the d subshell is

Table 3.2. Electron configuration for first-row d block atoms and ions in the gas phase [31, 41]. Higher charged ions have all been characterised as having a predictably decreasing number of d electrons. Two of the configurations are within $1000\,cm^{-1}$ of their first excited configuration: Ti^+ d^3 is $908\,cm^{-1}$ above the ground state; Ni d^9s^1 is only $205\,cm^{-1}$ above the ground state.

Ion	Sc	Ti	V	Cr	Mn	Fe	Co	Ni	Cu	Zn
0	d^1s^2	d^2s^2	d^3s^2	d^5s^1	d^5s^2	d^6s^2	d^7s^2	d^8s^2	$d^{10}s^1$	$d^{10}s^2$
+1	d^1s^1	d^2s^1	d^4	d^5	d^5s^1	d^6s^1	d^8	d^9	d^{10}	$d^{10}s^1$
+2	d^1	d^2	d^3	d^4	d^5	d^6	d^7	d^8	d^9	d^{10}

Table 3.3. Electron configuration for second-row d block atoms and ions in the gas phase [41, 42]. Where higher charged ions have been characterised, they have a predictably decreasing number of d electrons. The question mark indicates that the electron configuration hasn't been determined.

Ion	Y	Zr	Nb	Mo	Tc	Ru	Rh	Pd	Ag	Cd
0	d^1s^2	d^2s^2	d^4s^1	d^5s^1	d^5s^2	d^7s^1	d^8s^1	d^{10}	$d^{10}s^1$	$d^{10}s^2$
+1	s^2	d^2s^1	d^4	d^5	d^5s^1	d^7	d^8	d^9	d^{10}	$d^{10}s^1$
+2	d^1	d^2	d^3	d^4	?	d^6	d^7	d^8	d^9	d^{10}

Table 3.4. Electron configuration for third-row d block atoms and ions in the gas phase [42]. A question mark indicates that the electron configuration hasn't been determined. One of the configurations is within 1000 cm^{-1} of its first excited configuration: Pt d^8s^2 is 824 cm^{-1} above the ground state.

Ion	Lu	Hf	Ta	W	Re	Os	Ir	Pt	Au	Hg
0	d^1s^2	d^2s^2	d^3s^2	d^4s^2	d^5s^2	d^6s^2	d^7s^2	d^9s^1	$d^{10}s^1$	$d^{10}s^2$
+1	s^2	d^1s^2	d^3s^1	d^4s^1	d^5s^1	d^6s^1	?	?	d^{10}	$d^{10}s^1$
+2	s^1	?	?	?	?	?	?	?	?	d^{10}

full in the 2+ ions. This is consistent with the faster drop in energy of the d compared with the s subshell as atomic number increases, as shown in Figure 3.4.

The drop in the 3d subshell energy relative to 4s is made very clear by considering some isoelectronic series. Ca(s^2), Sc^+(d^1s^1), Ti^{2+}(d^2) illustrates this nicely, with later isoelectronic transition metal ions also adopting the d^2 configuration. Wheeler has added an interesting extra dimension with two isoelectronic series that include a negative ion [45]: Ca^-(s^2p^1), Sc(d^1s^2), Ti^+(d^2s^1), V^{2+}(d^3) and also Sc^-($d^1s^2p^1$), Ti(d^2s^2), V^+(d^4). Again, the later isoelectronic transition metal ions also adopt the d^n configuration. Not only do these series confirm the drop in the energy of the 3d subshell relative to the 4s across the period, but also they show that in the 1− ions up to scandium the valence p subshell is occupied. This shows the importance of the extra penetration of the 4p orbitals, which have two radial nodes, compared with the 3d, which don't have any. This confirms how orbital penetration is most important at the start of the period where there is the greatest shielding of the nucleus by the core electrons.

As the electrons are not independent in multi-electron atoms, observable energy changes in gas-phase atoms reflect differences in the *whole atom*,

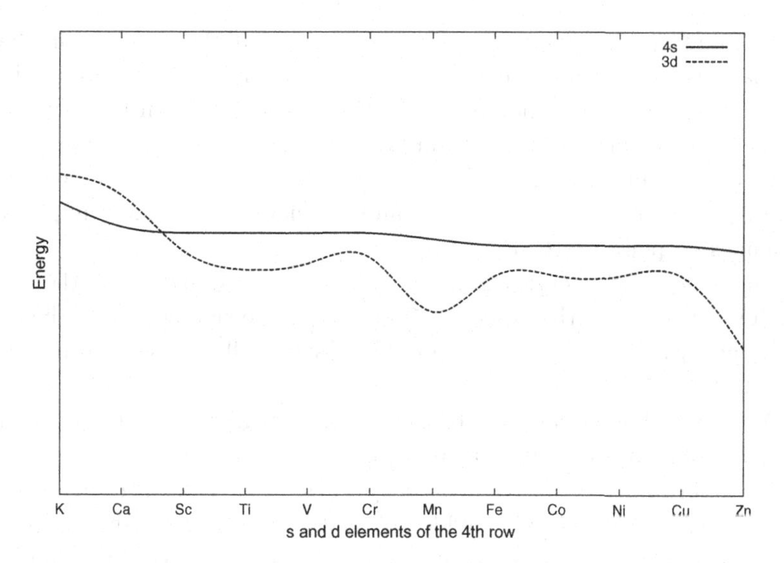

Figure 3.4. Schematic of 4s and 3d orbital energies for the s and d elements of the fourth row of the periodic table.

preventing direct measurement of orbital energies. For this reason, graphs of orbital energies in multi-electron atoms tend to be either schematic (with no numbers on the energy axis), e.g. [34], or showing a range of energies consistent with experiment, e.g. [52].

Keller has pointed out that these subshell energy diagrams are incapable of providing the numerical distribution of electrons between energy levels [53]. Vanquickenborne, referring to the 4s/3d subshell energy diagram with the levels crossing between Ca and Sc, writes that "this diagram is inadequate to discuss the orbital energy evolution of transition metals", owing to the sensitivity of the orbital energies for an element on its electron configuration [34]. Pilar has discussed how calculations of orbital energies in multi-electron atoms are commonly misinterpreted. He makes the point that individual electrons in many-electron atoms do not exist in energy levels, and that the word 'orbital' is commonly misused: the one-electron spin-orbital functions used in these calculations are not wavefunctions like the orbitals are in a hydrogen atom. Rather they are constructs of a model whose equations look a bit like the Schrödinger equation but they are not, and their associated quantum numbers do not refer to anything that is actually quantised. What is quantised is the overall wave function, which is not a product of the one-electron spin-orbitals, but rather a Slater determinant[3] (see Section 4.1). That is to say that the 'orbitals' in these calculations do not actually describe independent electrons, as is often assumed. Hence you cannot represent the promotion of an electron from one stationary state to another on these diagrams. He describes the use of these diagrams to describe 'orbitals' and 'energy levels' as if they were stationary quantum states as naive and conflicting with many important principles of quantum theory [54].

The schematic energies in Figure 3.4, contrived by the author, are roughly consistent with those of [34] and [52]. The significance of the dips in 3d energy at the half-full and full 3d subshell will be discussed in detail in Section 3.8.

An interesting explanation of the observed transition metal gas-phase electron configurations are the Rich and Suter diagrams, first proposed in 1988 [55] and recently promoted by Faria [56]. The diagrams, shown in Figures 3.5, 3.6 and 3.7, divide the energy for each subshell into two values, for up- and down-spinning electrons, represented by α and β, respectively. The diagrams show the s and d subshell energies decreasing with atomic number, with the d subshell energy decreasing more quickly, which is consistent with conventional subshell energy diagrams. The change in gradient in the dα line after d^5 is first reached in the 1st row diagram

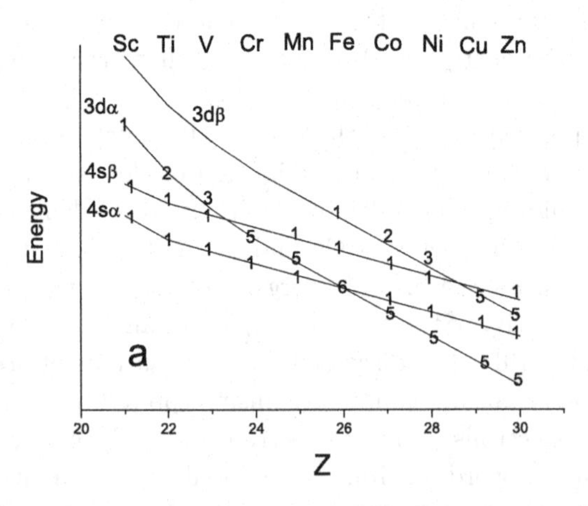

Figure 3.5. Rich and Suter diagrams for first-row d block metal atoms. (Figure produced and kindly provided with permission by Roberto Faria; adapted with permission from J. Chem. Ed., **65** (8), (1988), 702–4. ©1988 American Chemical Society.)

[3]Indeed, the full wavefunction is a sum of different Slater determinants.

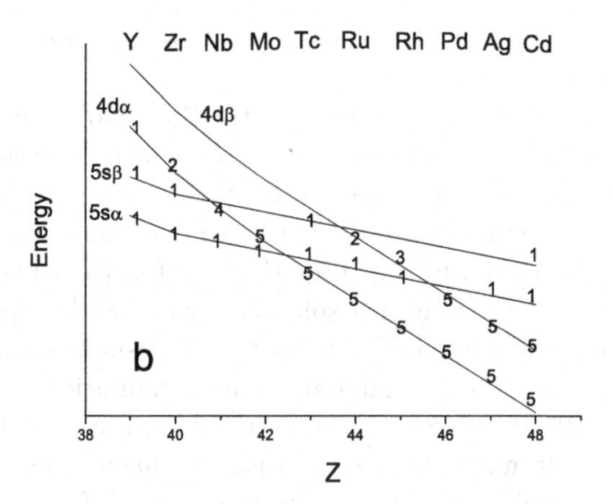

Figure 3.6. Rich and Suter diagrams for second-row d block metal atoms. (Figure produced and kindly provided with permission by Roberto Faria; adapted with permission from J. Chem. Ed., **65** (8), (1988), 702–4. ©1988 American Chemical Society.)

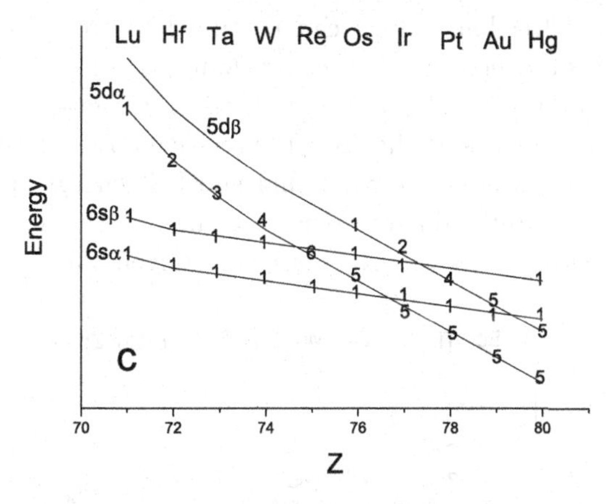

Figure 3.7. Rich and Suter diagrams for third-row d block metal atoms. (Figure produced and kindly provided with permission by Roberto Faria; adapted from Química Nova, **36** (6), (2013), 894–6, by permission of ©Sociedade Brasileira de Química.)

might be attributed to the end of the increase in exchange energy once d^5 is reached (see Section 3.8).

Two features of Rich and Suter diagrams are appealing. Firstly, they make explicit the energy cost in populating subshells beyond half-full — due to the additional repulsion of electrons that are paired in orbitals, together with the slowed increase of exchange energy compared with all-spin-parallel configurations (see Section 3.8). Secondly this approach can be made consistent with all the d block metal gas-phase electron configurations without exception.

Despite these pedagogical advantages, the Rich and Suter diagrams are not based on actual measurements or the sort of calculations that lead to the subshell energies seen in Figure 3.4. For this reason, they are not widely known. Their most serious failing is their placing of the valence ns subshell lower in energy than the $(n-1)$d for the first few elements in each row of the d block. Their notion of a spin-pairing energy is suggestive of an independent-electron approach based on the arrow-in-box microstate that is taken to represent the ground state of a configuration. This approach is overly simplistic compared to the actual energy calculations required for these atoms. It might be argued that the higher energy spin-down subshell energy level is, in effect, a representation of the lost exchange energy in more-than-half-full subshells compared with all-spin-parallel electron configurations. Rather than being absorbed in calculations, this manifestation of exchange energies in Rich and Suter diagrams effectively lends a degree of freedom, which can be tuned to create a scheme that correctly predicts all the d-block electron configurations.

While they can predict electron configurations for gas-phase atoms they do not work for ions without changing the relative positions of the subshell energies, as is also seen in conventional subshell energy diagrams. The modified Rich and Suter diagram that pertains to the 2+ ions of first-row d block metals is shown in Figure 3.8. Given the ambiguities and

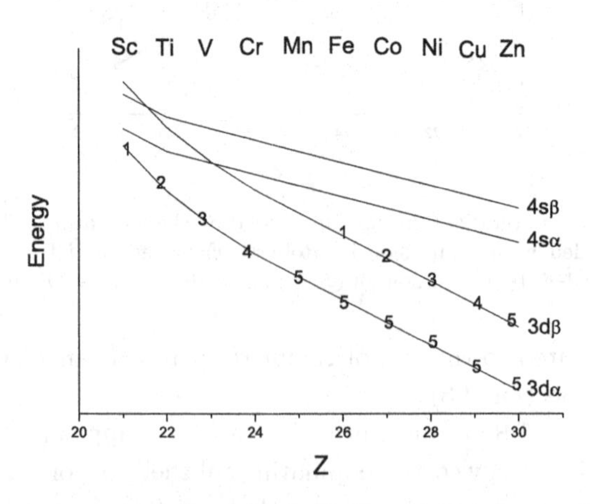

Figure 3.8. Rich and Suter diagram for 2+ first-row d block metal atoms. (Figure produced and kindly provided with permission by Roberto Faria; adapted from Química Nova, **36** (6), (2013), 894–6, by permission of ©Sociedade Brasileira de Química.)

misconceptions relating to conventional subshell energy diagrams, and the subsequent need to account for the s^1 and s^0 electron configurations, Rich and Suter diagrams possibly make for a more coherent introduction to transition metal electron configurations.

3.8 Irregular Configurations and Exchange Energy

As well as the classical direct-coulombic electrostatic repulsion between two electrons, there is a purely quantum mechanical interaction known as the exchange-coulombic interaction that is attractive but only operates between electrons of like spin. Its origins are in the antisymmetry of multi-electron wavefunctions and are explained fully in Section 4.2.

Exchange integrals can be used to explain the anomalous electron configurations observed in the gas-phase atoms of chromium ($4s^1\ 3d^5$) and copper ($4s^1\ 3d^{10}$). It is instructive to count the number of pairs of spin-parallel electrons that may be obtained from the configurations d^2 to d^{10}. In doing this, the spin-up and spin-down electrons need to be considered separately and we need to bear in mind that the electrons are indistinguishable. For either of these groups, the number of pairs of electrons that can be counted from a collection of n electrons is ${}^nC_2 = \frac{n!}{2!(n-2)!}$. The results of these calculations are shown in Table 3.5.

The $3d^5\ 4s^1$ configuration of chromium has ten possible pairs of parallel d electrons, while the $3d^4\ 4s^2$ configuration has only six possible such pairs. The energetic advantage due to exchange integrals from four more pairs of

Table 3.5. The number of pairs of parallel electrons in the ground states of the d configurations d^2 to d^{10}.

Configuration	Parallel electron pairs
d^2	1
d^3	3
d^4	6
d^5	10
d^6	10
d^7	11
d^8	13
d^9	16
d^{10}	20

parallel electrons is sufficient to promote an electron from the 4s subshell to the higher-energy 3d subshell. There is a similar energetic advantage of $3d^{10}$ $4s^1$ over $3d^9$ $4s^2$. Table 3.5 makes it clear that the exchange interactions for a subshell are maximised when the subshell is full. The greater exchange energy per d electron in d^5 and d^{10} compared with their neighbours explains the energy minima of the 3d subshell for these configurations in Figure 3.4. One cannot assume that the K exchange-coulombic integral will cause the same stabilisation energy in each d-block atom as the number of d electrons and protons is not constant; however, the K integrals should scale roughly with the direct-coulombic integrals, J, so the variation in K shouldn't negate this argument.

Another reason commonly given to explain the maximisation of parallel electrons is that it stabilises configurations through the avoidance of the additional coulombic repulsion experienced between two (spin-paired) electrons occupying the same orbital. Blake has explored the question of the relative importance of exchange-coulombic and direct-coulombic energies in maximising parallel electron spins. He considers the magnitudes of these integrals in terms of the radial Slater parameters (explained in Chapter 6) in spin-paired and spin-aligned configurations of p and of d electrons. His finding is that 70–80% of the effect in the d subshell is attributable to the exchange-coulombic energy, with the direct-coulombic energy responsible for the other 20–30%. Ranges are given as the exact value depends on which type of orbital (e.g. spherical harmonics or real linear combinations of them) are used in the calculation [57].

A very convincing justification of the important role played by the exchange interaction in subshell energies was made by P.G. Nelson in his PhD thesis (Cambridge University, 1962) and described by Johnson [58]. He considered the trend in third ionisation energy across the d block metals, i.e. removing a d electron from the $d^n s^0$ configuration, plotted in Figure 3.10 for the first- and second-row of the d block. This energy may be approximated to the energy of a d electron in the 2+ ion. The striking feature is the marked dip between manganese and iron in the first row and between technetium and ruthenium in the second row, within the otherwise steadily increasing trend. The third row is not plotted because of a lack of data.

The energy of the d electrons may be taken to result from three interactions: the electrostatic attraction to the core of the atom, E_{core}, taken to be the nucleus combined with the electrons of the nearest noble gas configuration; the direct-coulombic repulsion with the other d electrons, E_{dir}; and the attractive exchange-coulombic interaction with the other d electrons, E_{exch}.

$$E_{\text{total}} = E_{\text{core}} + E_{\text{dir}} + E_{\text{exch}}. \tag{3.8}$$

Since the d electrons do not shield each other effectively (see Section 3.3) each d electron's electrostatic attraction to the core increases roughly linearly as proton number, Z, and the number of d electrons, n, increase, tending to increase the ionisation energy. For *each* d electron the core electrostatic attraction is roughly

$$E_{\text{core}} = -(nU + c), \tag{3.9}$$

where U is the increased core attraction on the addition of a proton to the nucleus, c is a constant and the minus sign indicates that the energy is attractive. The direct-coulombic repulsion energy of all the d electrons is proportional to the number of electron pairs in the configuration, with a constant of proportionality, J:

$$E_{\text{dir}} = \frac{n!}{2!(n-2)!}J. \tag{3.10}$$

The attractive exchange-coulombic energy of all the d electrons is proportional to the number of pairs of electrons of like spin (α or β), with a constant of proportionality, K:

$$E_{\text{exch}} = -\left(\frac{n_\alpha!}{2!(n_\alpha-2)!} + \frac{n_\beta!}{2!(n_\beta-2)!}\right)K. \tag{3.11}$$

The most important of the three terms making up the d electron energy is the core attraction, which keeps the d electrons bound on the atom. With one fewer d electron after ionisation, the total core-attraction energy is less negative, leading to the ionised d^{n-1} species being higher in energy than d^n. The third ionisation energy, being a positive value, is found from $E_{\text{total}}(\mathrm{d}^{n-1}) - E_{\text{total}}(\mathrm{d}^n)$. Equation (3.9) needs to be multiplied by n to give the core-attraction energy of all the d electrons before ionisation, and by $n-1$ after ionisation. We use the result ${}^nC_2 - {}^{n-1}C_2 = n-1$ to give the following expression for the third ionisation energy:

$$E_{\text{total}}(\mathrm{d}^{n-1}) - E_{\text{total}}(\mathrm{d}^n) = nU + c - (n-1)J + (n_\alpha - 1)(n_\beta - 1)K. \tag{3.12}$$

With the core attraction being the most important term, the increasing nU outweighs the decreasing $-(n-1)J$, leading to the general increase in third ionisation energy that is observed. The third ionisation energy

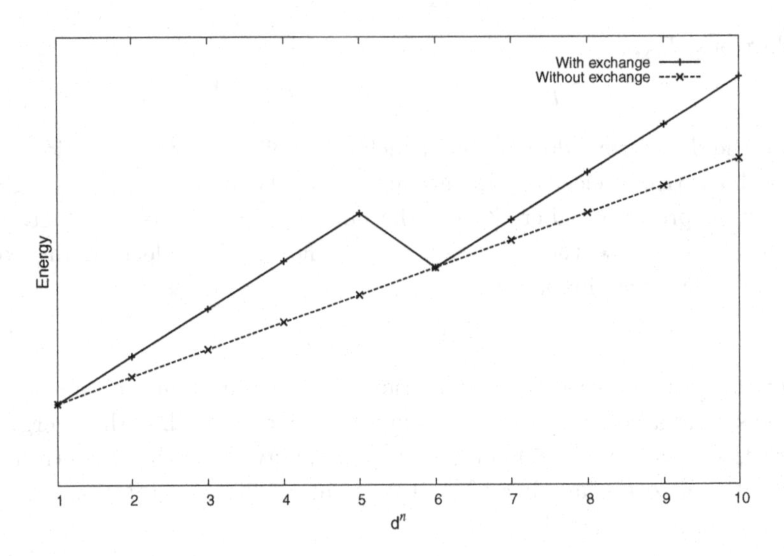

Figure 3.9. The Nelson model for third ionisation energies of d block elements with and without exchange energies [58].

of scandium is just given by $U + c$ as, with only a single d electron, there can be no direct-coulombic or exchange-coulombic energy within the d subshell.

Figure 3.9 shows a plot of third ionisation energies calculated from equation (3.12) against n (suitable values of U, c, J and K were chosen). The two lines in Figure 3.9 show the effect of the exchange energy when it is added to the core and repulsion terms. The shape of the plot has a striking resemblance to the plot of third ionisation energies measured for the first- and second-row d block elements in Figure 3.10 (third-row data are not available). This good agreement makes clear the importance of the exchange energy in trends in d orbital energy.

The maximum of exchange energy in full subshells is one of the reasons for the stability of noble gas configurations; another is the increasingly large energy jump up to the next vacant orbitals as the nuclear charge increases. It should however be stressed that the aufbau and exchange energy argument cannot account for several electron configurations in the gas-phase atoms of the second and third rows of the d block, shown in Table 3.6 [41, 42]. An interesting alternative approach to the trend in the relative energies of competing electron configurations across the three series in the d block is considered in the next section.

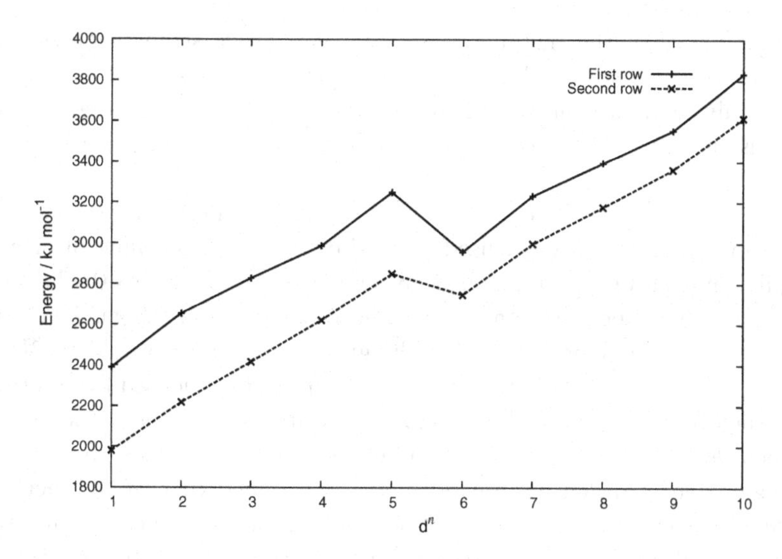

Figure 3.10. Observed 3rd ionisation energies for first- and second-row d block elements (data taken from [39]).

Table 3.6. Second and third row d block elements whose gas-phase electron configuration is inconsistent with those of the first row d block.

Metal atom	Predicted configuration	Actual configuration
Nb	$4d^3\ 5s^2$	$4d^4\ 5s^1$
Ru	$4d^6\ 5s^2$	$4d^7\ 5s^1$
Rh	$4d^7\ 5s^2$	$4d^8\ 5s^1$
Pd	$4d^8\ 5s^2$	$4d^{10}5s^0$
W	$5d^5\ 6s^1$	$5d^4\ 6s^2$
Pt	$5d^8\ 6s^2$	$5d^9\ 6s^1$

3.9 An Alternative Approach to d-Block Configurations

Until this point all discussion of electron configurations has been for gas-phase atoms. Two features of d-block gas-phase atoms make their electronic properties different from their metallic state and compounds.

Firstly, the large radial extent of the 4s orbital compared with other occupied orbitals means that in compounds it experiences Pauli repulsion with other atoms, raising its energy significantly. For example, in nickel tetracarbonyl, the nickel atom effectively has the electron configuration

$4s^0$ $3d^{10}$ [59]. In the metallic state, the 4s orbitals strongly overlap to form a very broad band that is only partially filled [59]. Hence in chemically relevant contexts, the outer s electrons have far less significance than is usually attributed to them as a result of considering gas-phase configurations.

Secondly, spin–orbit coupling can be significant in gas-phase atoms but it is not important in condensed phases, due to the quenching of orbital angular momentum. It turns out that spin–orbit coupling tips the balance between two competing configurations in nickel atoms in the gas phase. In the absence of spin–orbit coupling the ground-state term would be ^3D of $4s^1$ $3d^9$ but due to the greater spin–orbit coupling experienced by ^3F of $4s^2$ $3d^8$, the observed ground state in free nickel atoms is $^3F_4$ of $4s^2$ $3d^8$.

Schwarz has argued that the assignment of $4s^2$ $3d^8$ as the ground configuration of nickel is unrealistic. He has argued that in the metallic state the average energy of electron configurations is more chemically relevant, and has found the average energies for the $d^n s^0$, $d^{n-1}s^1$ and $d^{n-2}s^2$ configurations for the elements of the first three rows of the d block using data from the NIST Atomic Spectra Database [60]. He has filled gaps in the observed energy levels with calculated values [61].

Exercise 42

Work out the relative configuration-average energy, ΔE, of $3d^2$ $4s^1$ compared with $3d^1$ $4s^2$ in gas-phase scandium atoms. Use the data in Table 3.7 and quote the energy gap in eV to 3 significant figures. Consider $1\,\text{cm}^{-1}$ to be equivalent to $1.240 \times 10^{-4}\,\text{eV}$.

Table 3.7. The observed energy levels in the $3d^1$ $4s^2$ and $3d^2$ $4s^1$ configurations of Sc.

Config.	Term	J	E/cm^{-1}	Config.	Term	J	E/cm^{-1}
$3d^1$ $4s^2$	^2D	1.5	0	$3d^2$ $4s^1$	^2D	1.5	17025
$3d^1$ $4s^2$	^2D	2.5	168	$3d^2$ $4s^1$	^4P	0.5	17226
$3d^2$ $4s^1$	^4F	1.5	11520	$3d^2$ $4s^1$	^4P	1.5	17255
$3d^2$ $4s^1$	^4F	2.5	11558	$3d^2$ $4s^1$	^4P	2.5	17307
$3d^2$ $4s^1$	^4F	3.5	11610	$3d^2$ $4s^1$	^2G	4.5	20237
$3d^2$ $4s^1$	^4F	4.5	11677	$3d^2$ $4s^1$	^2G	3.5	20240
$3d^2$ $4s^1$	^2F	2.5	14926	$3d^2$ $4s^1$	^2P	0.5	20681
$3d^2$ $4s^1$	^2F	3.5	15042	$3d^2$ $4s^1$	^2P	1.5	20720
$3d^2$ $4s^1$	^2D	2.5	17013	$3d^2$ $4s^1$	^2S	0.5	26937

The configuration-average energies for the gas-phase atoms of the elements of the first three rows of d block elements found by Schwarz are shown in Figures 3.11, 3.12 and 3.13. The energies are given in eV relative to the $d^{g-1}s^1$ configuration for ease of presentation. Here, g is the number of electrons above the noble gas core in energy, or the group number of the element. Note that $1\,\text{eV} \approx 1.6 \times 10^{-19}$ J $\approx 96.5\,\text{kJ mol}^{-1} \approx 8056\,\text{cm}^{-1}$. Nowadays lutetium is regarded as a better candidate than lanthanum for group 3 in the sixth period on the basis of the periodicity of the elements' properties [62]. There is no data for the d^gs^0 configuration for group 11 elements (Cu, Ag, Au) as they have too many electrons for this configuration. Group 12 elements are not included as, with full s and d subshells, there is no competition from other configurations for the ground state. Spectroscopic data of the gas-phase atoms is sparse for the d^gs^0 configurations of the third-row d block elements, most of them not having a data point for this configuration in Figure 3.13, which will be well above the ground configuration anyway in cases where no data point is given. Osmium is the most problematic element: the $d^{g-1}s^1$ and $d^{g-2}s^2$ configurations are closer in energy than the uncertainties associated with those data points and so no confident assignment of the ground state can be made.

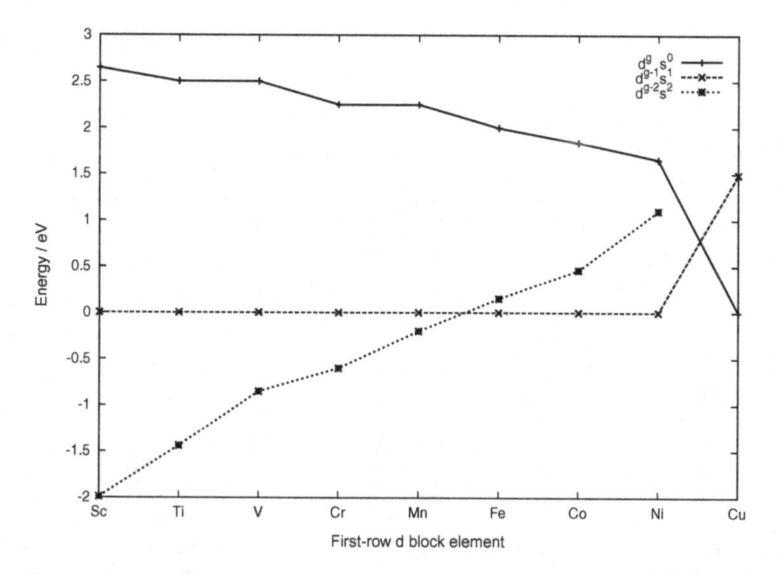

Figure 3.11. Gas-phase configuration-average energies for first-row d block metal atoms [61] (based on data kindly provided by the author). g is the number of electrons above the noble gas core in energy, or the group number of the element.

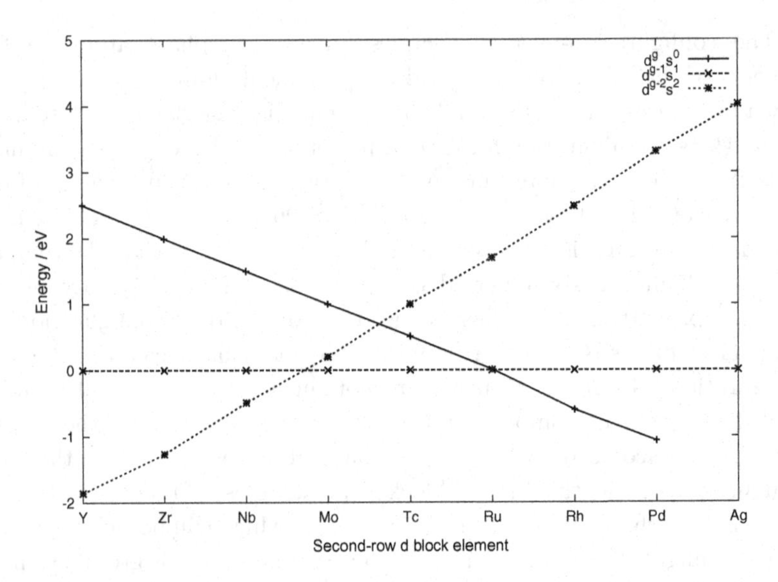

Figure 3.12. Gas-phase configuration-average energies for second-row d block metal atoms [61] (based on data kindly provided by the author). g is the number of electrons above the noble gas core in energy, or the group number of the element.

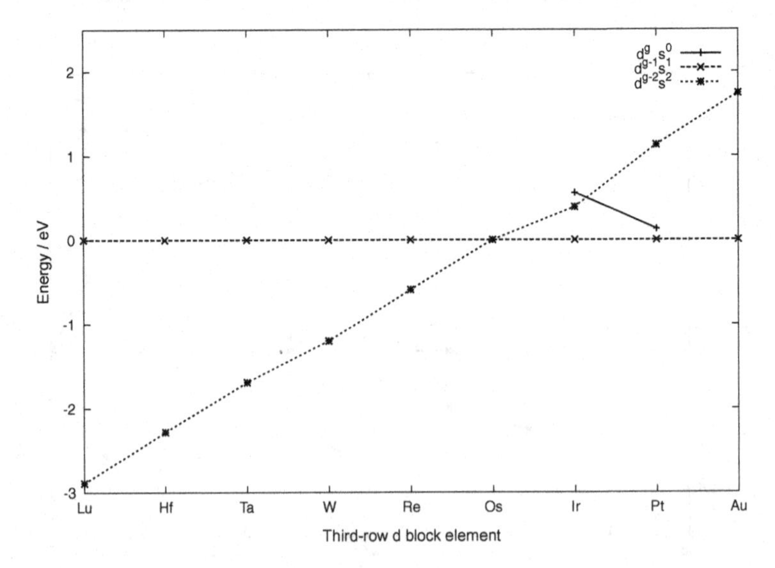

Figure 3.13. Gas-phase configuration-average energies for third-row d block metal atoms [61] (based on data kindly provided by the author). g is the number of electrons above the noble gas core in energy, or the group number of the element.

Table 3.8. The configurations of lowest average energy for the gas-phase atoms of the elements of the first three rows of the d block (except group 12). Note that g is the total number of electrons above the noble gas core in energy, or the group number of the element.

Row	$d^{g-2}s^2$	$d^{g-1}s^1$	$d^g s^0$
1st	Sc Ti V Cr Mn	Fe Co Ni Cu	—
2nd	Y Zr Nb	Mo Tc Ru Rh Pd Ag	Rh Pd
3rd	Lu Hf Ta W Re Os	Os Ir Pt Au	—

Table 3.9. Elements whose configuration-average and ground-state electron configurations are different. Note that g is the total number of electrons above the noble gas core in energy, or the group number of the element.

d block metal	Ground-state	Configuration-average
Cr Nb	$d^{g-1}s^1$	$d^{g-2}s^2$
Fe Co Ni Tc Os Ir	$d^{g-2}s^2$	$d^{g-1}s^1$
Rh	$d^{g-1}s^1$	$d^g s^0$

In summary, the configurations of lowest average energy for the gas-phase atoms of the elements of the first three rows of the d block are given in Table 3.8. d block elements whose configuration-average and ground-state electron configurations are different are collected in Table 3.9.

What is appealing about this approach is that the configuration-average energies in all three rows of the d block vary in a more-or-less regular way, and the relative energies between the different types of configuration vary in the same sense in all three rows, i.e. with configurations having greatest d-subshell occupancy becoming relatively more favoured towards the right of each row in the d block. This is consistent with their core-like nature given the increasing effective nuclear charge across the period. In this sense the description of electron configurations in this way is less complicated by exceptions than is the traditional approach.

The variation between the three periods is not a regular one, however. Given the lack of regularity in the chemical and physical properties of the three rows this isn't surprising. The first row may be regarded as special in the sense that it is the only one where the d electron wavefunctions are radially nodeless, accentuating the electron–electron repulsion experienced by the d electrons. The third row is regarded as special in the sense that it is the only one of the three with an inner f shell, and hence a lanthanide contraction. It is in the third row, therefore, that relativistic effects are prominent. Relativistic effects and the lanthanide contraction will be discussed in Section 3.10. Curiously then, one might conclude that the second row of the d block is the only 'normal' one. Similar arguments have been used to account for the fact that in group 11 (ignoring the fourth row) silver is the only metal that isn't coloured.

Schwarz's approach of plotting the energy of whole configurations is reminiscent of the approach taken by Kuhn in his 1962 monograph on atomic spectra [63] — they both plot the energy of $d^{g-2}s^2$ and $d^g s^0$ relative to that of $d^{g-1}s^1$. The difference between them is that Kuhn took the energy of a configuration to be its lowest energy state (actually, the highest spin-multiplicity state but this will tend to be the lowest energy state according to Hund's rules, which are discussed in Section 4.6) while Schwarz used the configuration's average energy. Interestingly, it is only Schwarz's approach that produces smooth changes in all the configuration energies across each series of d block elements. There are two advantages of their whole-configuration approaches over subshell energy diagrams. Firstly, they are clearly based on experimental data, making their meaning transparent — in contrast to the subshell energy diagrams that rely on complicated calculations whose assumptions are far from transparent. Secondly, by considering whole configurations, there is less suggestion that electrons may be treated independently than there is in subshell energy diagrams.

Schwarz has also proposed a more flexible approach to the aufbau ordering of subshells: that there should be multiple schemes depending on the region of periodic table being considered [59]. Obviously, hydrogen is a special case with all subshells within a principal shell being degenerate. Schwarz suggests that the conventional aufbau ordering given in Section 3.2 should just be reserved for the s block, for which the nd subshell is higher in energy than the $(n+1)$s. For most of the rest of the periodic table, the following scheme is a realistic ordering for valence shells:

$$\begin{gathered}1s \ll 2s < 2p \ll 3s < 3p \ll 3d < 4s < 4p \ll 4d < 5s < 5p \ll \\ 4f < 5d < 6s < 6p \ll 5f < 6d < 7s < 7p \ldots\end{gathered}$$

The $\ll$ symbols appear after p subshells (and also 1s) and relate to the jump in energy after a noble gas configuration. Note that the d (and f) subshells that are conventionally seen after 4s and higher-s subshells appear before it (with f appearing before d). This is in accordance with these inner subshell energies dropping below the valence s subshell at the start of their respective series (and f subshells dropping in energy more rapidly than d).

Another ordering of subshell energies that may be found are those within the core of very heavy atoms, after the d and f subshells have dropped far enough in energy so that they are ordered with the other subshells of their principal shell:

$$1s \ll 2s < 2p \ll 3s < 3p < 3d \ll 4s < 4p < 4d < 4f \ll \ldots$$

Within the core of very heavy atoms the large energy gaps between subshells appear after the principal shells are filled, rather than after noble gas configurations. By the seventh period of the periodic table, the energy of the 4f subshell is below that of the 5s, according to Figure 8.3 in [50]. Orderings that are intermediate between these core and the above valence orderings are also conceivable.

3.10 Explaining Some Surprising d-Block Chemistry

First row

We saw in Section 3.8 that, owing to the exchange energies in the d subshell, the third ionisation energy of manganese is significantly higher than it is for its neighbour, iron, and more similar to that of cobalt than any other 3d metal (see Figure 3.10). This makes sense of the surprising observation that Mn(III) and Co(III) are powerfully oxidising while Fe(III), between them, is only mildly oxidising.

Chromium(III) d^3 exhibits some surprising chemistry: it is more amphoteric than other 3+ ions from the first row of the d block — its hydroxide is the only one that readily dissolves in NaOH(aq) — and it is extraordinarily kinetically inert. Indeed, the rate of exchange of water ligands on Cr^{3+}(aq) is over seven orders of magnitude slower than on the next slowest ion from the first row of the d block, which is V^{2+}, also d^3 [64]. First-row d block ions form 'high spin' octahedral hexaaquacomplexes in water. The degeneracy of the five d orbitals is broken in octahedral symmetry into two groups: the three orbitals that point between the Cartesian axes, given the symmetry label t_{2g}, and the two orbitals that point along the Cartesian axes, given the label e_g. 'High spin' indicates that after the three lower-energy t_{2g} orbitals are singly occupied, the fourth d electron preferentially occupies one of the higher-energy e_g orbitals rather than pairing up with an electron in one of the t_{2g} orbitals (exchange energy promotes this configuration). Following interaction with orbitals on the coordinating water molecules, the e_g orbitals become anti-bonding, while the t_{2g} orbitals remain largely non-bonding. The d^3 configuration is special since the exchange energy per electron is maximised in the non-bonding orbitals, while the anti-bonding orbitals are vacant. This leads to a stabilisation of the complex that accounts for its inert ligand exchange. The strong coordinate bonds that result from interaction between the d^3 ion and the water ligands — stronger in Cr^{3+} than V^{2+} due to its smaller size and the stronger ion–dipole interaction with a 3+ charge — account for the induced acidity in the water ligand molecules that creates the amphoteric properties.

Aqueous transition metal ions, i.e. those with a partially occupied d subshell, are known for being coloured. It is surprising then that Mn^{2+} d^5 is nearly colourless, and then troubling that Fe^{3+}, also d^5, is quite strongly coloured. The pre-university explanation for colour in transition metal ions is the absorption of visible light leading to a d–d transition, exciting a d electron from the lower t_{2g} level to the upper e_g level. In Mn^{2+}(aq), being high-spin, all five d orbitals are singly occupied. The arrow-in-box notation would describe the configuration with five up-spinning electrons, one in each orbital. A more sophisticated treatment (see Sections 4.3 and 4.6) would describe it as having the maximum spin multiplicity. When light excites an electron to a higher energy level, it is principally through the oscillating electric field of the radiation, which has no influence on the spin of the electrons (except for heavy elements which are more complicated). This accounts for the d^5 Mn^{2+} ion being nearly colourless since there is no vacancy in the e_g level for an electron of the same spin as the two already in there.

Fe^{3+} d^5 gets its colour from a different mechanism. The electronic excitation is from an orbital essentially on a water ligand to an orbital essentially on the metal; this is known as a ligand-to-metal charge-transfer transition (LMCT). This is a feature of higher oxidation state transition metal ions, which polarise their ligands — most famously manganate(VII), chromate(VI) and vanadate(V). A d–d transition is not possible for these three examples, since they are all d^0 configurations.

Across the first-row d block metals, the trends in atomisation energy, melting point and boiling point all have a double-hump appearance, with a minimum at manganese and maxima either side of it. The double-hump is a sum of two parabolic curves spanning the series: one with a maximum at manganese and the other with its minimum at manganese. The first represents the extent of the metallic bonding interaction from the overlap of d orbitals on adjacent orbitals, which would be expected to be at a maximum at the half-filled subshell. The curve with a minimum at manganese gives the drop in energy on the free atoms following the reorganisation of the d electrons to maximise the number of parallel electrons, i.e. it is a measure of the exchange energy. This is explained further on pages 176–7 of [58].

Second row

There is a significant change in the redox properties when passing from the first to the second row of the d block. The third-row elements are relatively similar to the second-row ones in this regard. Higher oxidation states are generally more stable in the second and third rows, especially in the middle of the series. This can be related to the radial form of the d orbitals in each row of the d block. The 3d orbitals are relatively core-like with respect to the s and p orbitals in their shell and therefore relatively difficult to ionise and to involve in bonding interactions. The 4d and 5d orbitals have a greater radial extent relative to their respective s and p orbitals, making them more available for bonding and susceptible to ionisation. The 3d orbitals are the only d orbitals with no radial node, which makes them more core-like.

The radial density functions, $r^2\psi(r)^2$, of the 3s, 3p and 3d orbitals of V^{4+} are plotted in Figure 3.14; the radial density functions of the 4s, 4p and 4d orbitals of Nb^{4+} are plotted in Figure 3.15 for comparison. We need to correct Z to take the shielding of these electrons into account using Slater's rules (see Section 3.3). We write the electron configuration of V^{4+} as $[_2He]$ $(2s\ 2p)^8$ $(3s\ 3p)^8$ $3d^1$. The shielding factor, S, for the 3d electron is 18; for the 3s and 3p electrons it is $(0.35 \times 7) + (0.85 \times 8) + (1.00 \times 2) = 11.25$. This gives $Z_{\text{eff}}(3d) = 23 - 18 = 5$ and $Z_{\text{eff}}(3sp) = 23 - 11.25 = 11.75$. We write the electron configuration of Nb^{4+} as $[_{10}Ne]$ $(3s\ 3p)^8$ $3d^{10}$ $(4s\ 4p)^8$ $4d^1$. The shielding factor, S, for the 4d electron is 36; for the 4s and 4p electrons it is $(0.35 \times 7) + (0.85 \times 18) + (1.00 \times 10) = 27.75$. This gives $Z_{\text{eff}}(4d) = 41 - 36 = 5$ and $Z_{\text{eff}}(4sp) = 41 - 27.75 = 13.25$. Figures 3.14 and 3.15 make it clear that 3d orbital is significantly more core-like than the 4d orbital, accounting for the greater reactivity of the d electrons in the second row of the d block compared with the first.

The maximum in the 3d radial density function in Figure 3.14 is in agreement with the 1.8 a_0 predicted by the n^2a_0/Z_{eff} of equation (2.32), and is lower than the expectation value, $\langle r \rangle$, shown in Table 3.10, due to the asymmetric shape of the 3d radial density function.

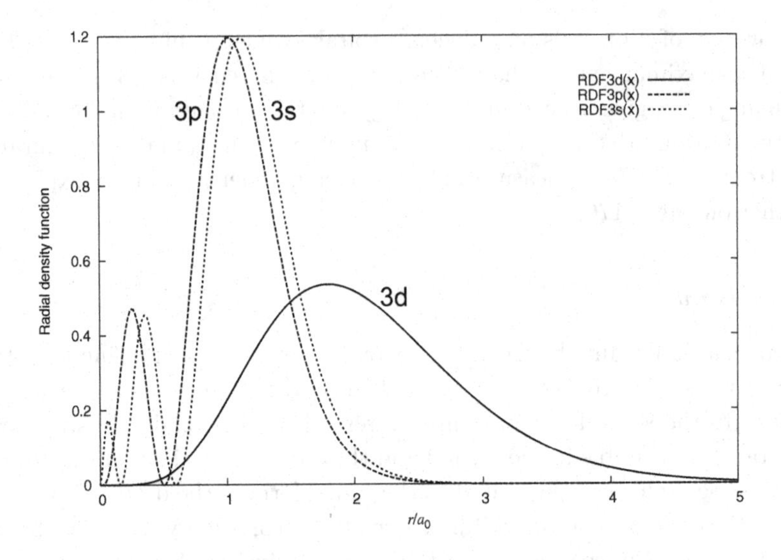

Figure 3.14. The radial density functions of the 3s, 3p and 3d orbitals of V^{4+}.

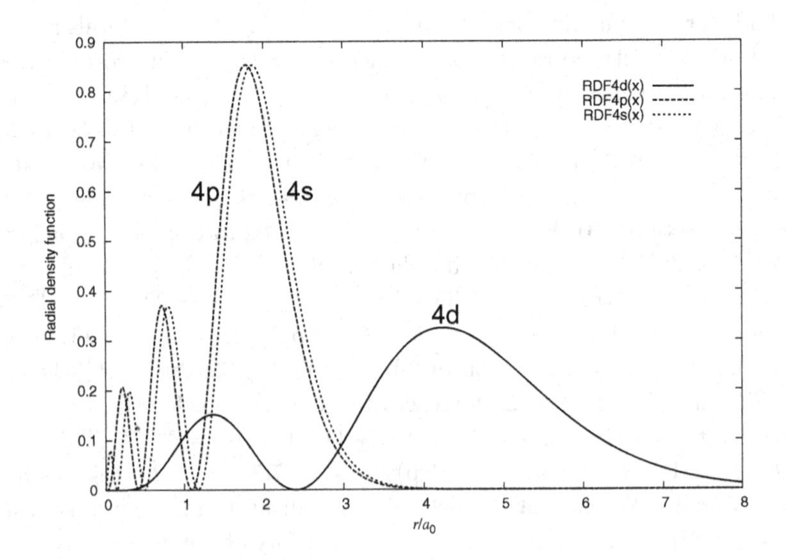

Figure 3.15. The radial density functions of the 4s, 4p and 4d orbitals of Nb^{4+}.

Table 3.10. The expectation values, $\langle r \rangle$, in Bohr radii, a_0, for the electrons in the outer d, s and p subshells in the V^{4+} and Nb^{4+} ions, the weighted average of the s and p values, and a comparison between the two ions, following Slater's rules.

Metal ion	$\langle r_d \rangle$	$\langle r_s \rangle$	$\langle r_p \rangle$	$\frac{1}{4}(\langle r_s \rangle + 3\langle r_p \rangle)$
V^{4+}	2.10	1.15	1.06	1.09
Nb^{4+}	4.20	1.81	1.74	1.75
Nb^{4+}/V^{4+}	2.00	1.58	1.63	1.62

The expectation values, $\langle r \rangle$, for the electrons in the orbitals in Figures 3.14 and 3.15 are shown in Table 3.10 using the Z_{eff} values with equation (2.42).

The numbers in the table appear consistent with the plots of the radial density functions in Figures 3.14 and 3.15. The ratios of the d expectation value to the weighted-average s and p expectation values are 1.94 for the first-row element and 2.39 for the second-row element. This is consistent with the 4d orbitals having a greater radial extent relative to the other valence subshells than do the 3d orbitals. Energetic factors will of course be crucial in accounting for the different chemistry between the first and second rows of the d block.

The decreasingly core-like nature of the outer d orbitals compared with the outer s and p orbitals on descending the d block might have been predicted by inspection of the expectation value equation (2.42). If we take the outer orbitals to be s orbitals, whose $\langle r \rangle$ is a good approximation to the value for the p orbitals in the same shell, then, on increasing n, the $\langle r \rangle$ will vary roughly as $3n^2/2$ as Z_{eff} is likely to remain constant as additional core shells of electrons negate the additional attraction from the additional protons in the nucleus. For higher subshells though, where the final term in equation (2.42) is non-zero, the increasing value of n will serve to increase the value of $\langle r \rangle$ at a faster rate. It is perhaps also significant that as the value of n increases for d orbitals, the most probable distance of the electron from the nucleus becomes greater than the expectation value on account of the increasing number of radial nodes in the wavefunction; for the 3d orbitals, which have no radial nodes, the most probable distance is smaller than the expectation value.

Third row

At the end of Section 3.9, reference was made to relativistic effects and the lanthanide contraction causing anomalies in the third row of the d block. The lanthanide contraction describes how the atomic radii of the third-row d block are scarcely larger than those of the second row, despite the second-row atoms being significantly larger than their first-row neighbours. This reflects the increase of proton number from the second to third row being 32 rather than 18, due to the extra 14 lanthanide elements intervening between the 6s and 5d elements. With the jump in atomic number to the third row, relativistic effects, discussed in Section 2.10, start to become visible. The heaviest element in the third row, mercury with $Z = 80$, will have extremely fast-moving electrons. Assuming the 1s electrons experience nearly all of the nuclear charge their speed will be over half the speed of light. When $v = c/2$, the relativistic gamma term of equation (2.46) equals $\frac{2}{\sqrt{3}}$, which causes a 15% increase in the mass of the electron, and therefore a 15% decrease in the Bohr radius, a_0, of equation (2.26). This is known as the heavy fermion effect and causes the affected electrons to be more core-like. The penetration of 1s electrons into the core of the atom and the finite value of their wavefunction at the nucleus itself makes it particularly susceptible to this effect. It is significant that the contraction of the 1s orbital's Bohr radius must also occur for all the higher s orbitals on account of their mutual orthogonality. This requirement does not, however, extend to other subshells — again, due to orthogonality arguments.

Exercise 43

Show how the orthogonality of the different s orbitals requires that the reduction in the 1s Bohr radius applies to the higher s orbitals as well.

Exercise 44

Explain why mutual orthogonality does not require subshells other than s to be contracted to the same extent as the 1s orbital.

Following this contraction, the more core-like 6s orbital in mercury participates less in metallic bonding, and this is thought to account for mercury being a liquid at room temperature. The $d^{10}s^2$ configuration with its full subshells leads to weak metallic bonding in any case, as shown by zinc and cadmium in its group, but the drop in melting and boiling point from cadmium to mercury is far larger than the drop from zinc to cadmium.

Recent calculations show that the relativistic contraction of the 6s orbital in mercury causes a 105 K drop in its melting point [65], which is sufficient to make the difference between mercury being a liquid or a solid at room temperature.

The element beneath mercury in the d block is the recently synthesised copernicium, Cn, officially recognised by IUPAC in 2009. It is thought to have such a relativistically contracted 7s orbital that it is a gas at room temperature and pressure [66]. Indeed, it has been calculated that it will ionise its 6d electrons before the 7s [67] (leading to a partially full d subshell, making it a genuine transition element unlike the other members of its group).

With the outermost s and d subshells being so close in energy in d block elements, a relativistic contraction of the s orbitals would be expected to have a significant impact on the chemistry of the heaviest d block elements. Indeed the other element with the most well-known relativistic effects is mercury's neighbour gold, whose relativistic properties have been studied in great detail by Pyykkö [68].

The relativistic effects on the elements following mercury at the bottom of the p block might be expected to be less significant, as the 6p subshell will be the most important one in bonding interactions and will not experience the same effect. However, the inert pair effect in these elements is a manifestation of the relativistic contraction of the 6s orbital, since the 6s electrons become less available for bonding and harder to ionise, stabilising the valency two less than the number of s and p valence electrons in the neutral atom.

3.11 Contrasts in f-Block Chemistry

The observed electron configurations of the gas-phase lanthanide (4f) elements and their first few ions are shown in Table 3.11. There are some common features with the d block, e.g. the valence s orbital is occupied across the series, despite being at higher energy than the inner subshell — 4f in the case of the lanthanides. Similarly, the valence s electrons are ionised first and the most core-like subshell (4f in this case) last. The 5d subshell is only occupied early in the lanthanide series, but appears again at gadolinium so that it can have an f^7 configuration, promoted by the additional exchange energy to be gained by the 5d electron spinning parallel with the half-filled 4f subshell, rather than there being an additional 4f electron spinning anti-parallel. The 5d subshell is intermediate in energy

Table 3.11. Electron configuration for lanthanide (4f) atoms and ions in the gas phase [69]. Where higher charged ions have been characterised, they have a predictably decreasing number of f electrons. One of the configurations is within 1000 cm^{-1} of its first excited configuration: Tb $f^8d^1s^2$ is 285 cm^{-1} above the ground state.

Ion	La	Ce	Pr	Nd	Pm	Sm	Eu
0	d^1s^2	$f^1d^1s^2$	f^3s^2	f^4s^2	f^5s^2	f^6s^2	f^7s^2
+1	d^2	f^1d^2	f^3s^1	f^4s^1	f^5s^1	f^6s^1	f^7s^1
+2	d^1	f^2	f^3	f^4	f^5	f^6	f^7
+3		f^1	f^2	?	f^4	f^5	f^6
Ion	**Gd**	**Tb**	**Dy**	**Ho**	**Er**	**Tm**	**Yb**
0	$f^7d^1s^2$	f^9s^2	$f^{10}s^2$	$f^{11}s^2$	$f^{12}s^2$	$f^{13}s^2$	$f^{14}s^2$
+1	$f^7d^1s^1$	f^9s^1	$f^{10}s^1$	$f^{11}s^1$	$f^{12}s^1$	$f^{13}s^1$	$f^{14}s^1$
+2	f^7d^1	f^9	f^{10}	f^{11}	f^{12}	f^{13}	f^{14}
+3	f^7	f^8	f^9	f^{10}	f^{11}	f^{12}	f^{13}

between the 6s and 4f subshells, but does not provide the far more radially extended orbital of the 6s subshell, which is why it is less often occupied to stabilise the 4f configuration.

The 4f subshell is not occupied at all in the gas-phase ground state of lanthanum or its ions, which is often therefore considered a group 3 element rather than a 4f one. However, the 4f subshell is fully occupied at ytterbium, which is why the next element, lutetium, which is never chemically observed to have a partially-full 4f subshell, is increasingly these days considered the first element of the third row of the d block [62]. Its electron configuration, [Xe] $4f^{14}$ $5d^1$ $6s^2$, is more similar to the next element, hafnium [Xe] $4f^{14}$ $5d^2$ $6s^2$, than is lanthanum. With lutetium in the d block, and with seven f orbitals implying a 14-element series, lanthanum then becomes the first lanthanide (as its name might suggest).

The observed electron configurations of the gas-phase actinide (5f) elements are shown in Table 3.12 (the data for the gas-phase ions are not known). Like the lanthanides, each atom has a full valence s subshell in the gas phase. Again, the valence d subshell is occupied in early members of the series and also just after the half-filled f subshell has been reached to enable a second element to have an f^7 configuration (favoured by exchange energies, as before). In keeping with the approach to the lanthanides, the actinides are increasingly these days taken to be the 14 elements immediately following the s block.

While Rich and Suter diagrams can account for the electron configuration of every d block element, they are less successful in the f block. Of the

Table 3.12. Electron configuration for actinide (5f) gas-phase atoms [70].

Ac	Th	Pa	U	Np	Pu	Am
d^1s^2	d^2s^2	$f^2d^1s^2$	$f^3d^1s^2$	$f^4d^1s^2$	f^6s^2	f^7s^2
Cm	Bk	Cf	Es	Fm	Md	No
$f^7d^1s^2$	f^9s^2	$f^{10}s^2$	$f^{11}s^2$	$f^{12}s^2$	$f^{13}s^2$	$f^{14}s^2$

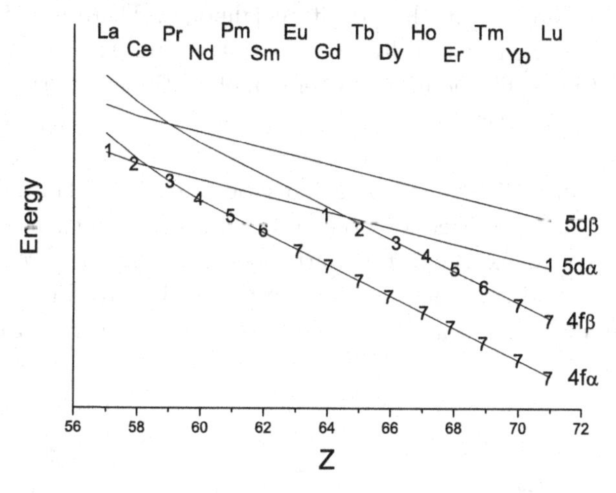

Figure 3.16. Rich and Suter diagram for lanthanide atoms. (Figure produced and kindly provided with permission by Roberto Faria; adapted from Química Nova, **36** (6), (2013), 894–6, by permission of ©Sociedade Brasileira de Química.)

actinides, there is no way they can account for the configurations of Pa, U and Np on account of their valence d and f subshells being simultaneously less than half-full. The Rich and Suter diagram for the lanthanides is shown in Figure 3.16. Cerium ($f^1d^1s^2$) is similarly impossible for the diagram to predict. These anomalies are the result of either a very fine energy balance between the valence d and f subshells or interactions between competing configurations mixing them together.

An exceptional property of the lanthanides is that they are chemically very similar, with most of their chemistry limited to the +3 oxidation state (this is why they are so difficult to separate). In a similar way to the d block elements they ionise the electrons from the outer shells first, which means $6s^2$ $5d^1$ for the early lanthanides and gadolinium; the 4f subshell is more core-like and remains largely inert once the 3+ oxidation state is reached —

in contrast to the d block elements, whose partially full d subshells interact with ligands or may be further ionised readily.

The observed third ionisation energies of the lanthanides are plotted in Figure 3.17. The difference in exchange energies that produce the trend in ionisation energies observed for d block elements in Figure 3.10 also applies to the lanthanides: a general increase across the series with a marked drop after the half-filled subshell. This leads to an analogous chemical consequence: Eu^{2+}, being at a maximum, has greater stability compared with its 3+ ion than do its neighbours. There are three other lanthanides with well-known divalent chemistry: Sm, Tm and Yb. This is consistent with the third ionisation energy plot, as these have the next three highest values after Eu.

The observed fourth ionisation energies of the lanthanides are also plotted in Figure 3.17. The fourth ionisation energy of lanthanum isn't plotted as it is much higher than the others, since it is the xenon core that is being ionised. As expected, the fourth ionisation energies are all greater than the third, and the trend is displaced to the right by an element since there is one fewer electron. Tetravalent chemistry is most well known for

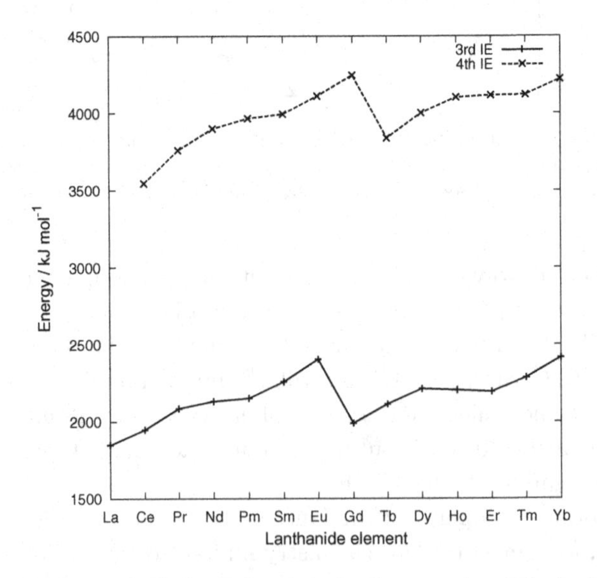

Figure 3.17. The observed third and fourth ionisation energies of the lanthanides from [69], applying the conversion factor $1\,\text{eV} = 96.485\,\text{kJ mol}^{-1}$. The fourth ionisation energy of lanthanum isn't plotted as it is much higher than the others, since it is the xenon core that is being ionised.

cerium, and the two next most well-known tetravalent lanthanides are Pr and Tb, which is entirely consistent with this data.

A curiosity of the lanthanide ionisation energy trends is that there are also discontinuities at the quarter- and three-quarter-filled subshell, particularly the latter, which cannot be explained by exchange energies or the general arguments applied in Section 3.8. They are more evident in the third ionisation energy plot but are still discernible at the next level of ionisation. This too has chemical consequences: lanthanide dihalides are also known for Nd^{2+} f^4 and Dy^{2+} f^{10}.

The origin of the quarter- and three-quarter-filled subshell effects is the projection of the orbital angular momentum of the f electrons on to the principal axis. In the arrow-in-box scheme, when the number of parallel f electrons is less than or equal to $\frac{1}{2}(l+1)$ all the electrons may occupy orbitals of non-negative m. Negative values of m imply rotation of the electrons around the z-axis in the opposite sense. When rotation of the electrons around the axis is in the same sense the electrons can most effectively avoid one another, and this reduces the electrostatic repulsion between them — in accordance with Hund's rules of Section 4.6. The extra repulsion from the occupation of a negative m orbital accounts for the discontinuity in the third ionisation after the quarter- and three-quarter-filled f subshell. This effect may be most significant for the f elements as this block of the periodic table is where orbital angular momentum is greatest. Further discussion may be found in pages 158–168 of [58]. A lack of ionisation data precludes such an analysis of the actinides.

Figure 3.18 shows the radial density functions for the 4f, 5s and 5p orbitals in trivalent neodymium, Nd^{3+}. We write the electron configuration of the ion as $[_{18}Ar]$ $3d^{10}$ $(4s\ 4p)^8$ $4d^{10}$ $4f^3$ $(5s\ 5p)^8$. Z_{eff} is calculated separately for the fourth- and fifth-shell orbitals, as they will have different shielding factors, S (see Section 3.3). Following Slater's rules, we find that for 4f electrons, $S_{4f} = (0.35 \times 2) + (1.00 \times 46) = 46.70$, while for the 5s and 5p electrons, $S_{5sp} = (0.35 \times 7) + (0.85, \times 21) + (1.00 \times 28) = 48.30$. Since $Z = 60$ for Nd we take Z_{eff} to be 13.30 for 4f electrons and 11.70 for 5s and 5p electrons. Following remarks in Section 3.7 about the effect of electron–electron interactions in the valence shell, we can see that our present approach is only a rough approximation. Nevertheless, Figure 3.18 gives a clear picture of how the 4f electrons are shielded from the local chemical environment by the electrons of the fifth shell, showing why the 4f subshell is often considered to be core-like. In the calculation of the

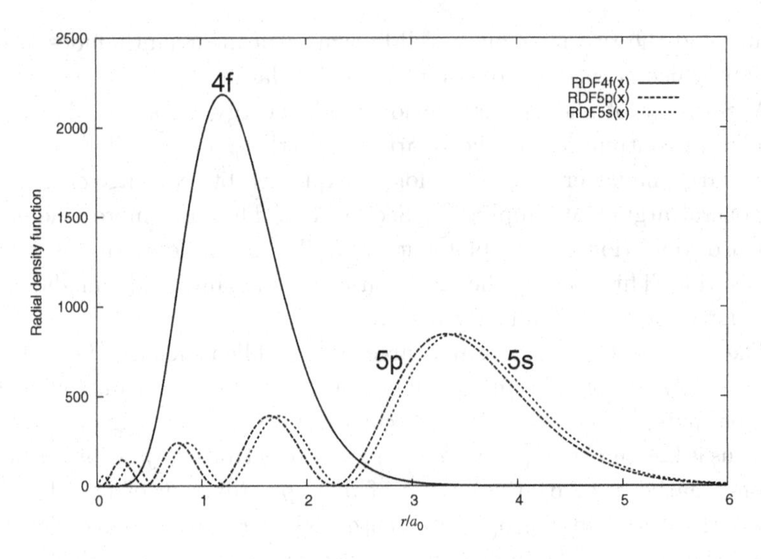

Figure 3.18. Radial density functions for the 4f, 5s and 5p subshells in Nd^{3+} using Slater's rules.

Table 3.13. The expectation values, $\langle r \rangle$, in Bohr radii, a_0, for the electrons in the outer f, s and p subshells in the Nd^{3+} and U^{3+} ions, the weighted average of the s and p values, and a comparison between the two ions, following Slater's rules.

Metal ion	$\langle r_f \rangle$	$\langle r_s \rangle$	$\langle r_p \rangle$	$\frac{1}{4}(\langle r_s \rangle + 3\langle r_p \rangle)$
Nd^{3+}	1.35	3.21	3.12	3.14
U^{3+}	2.37	4.62	4.53	4.55
U^{3+}/Nd^{3+}	1.75	1.44	1.45	1.45

radial density functions Z is replaced by Z_{eff} but n is not modified (see Section 3.3).

The maximum in the 4f radial density function in Figure 3.18 is in agreement with the $1.20\,a_0$ predicted by the $n^2 a_0/Z_{\text{eff}}$ of equation (2.32), and is lower than the expectation value, $\langle r \rangle$, shown in Table 3.13, due to the asymmetric shape of the 4f radial density function.

The 5f electrons in the actinides are less core-like than the 4f lanthanide electrons, and so more readily participate in bonding interactions and are easier to ionise. Neodymium's sister element in the actinide series, uranium ($Z = 92$), forms stable hexavalent compounds, for example. Figure 3.19

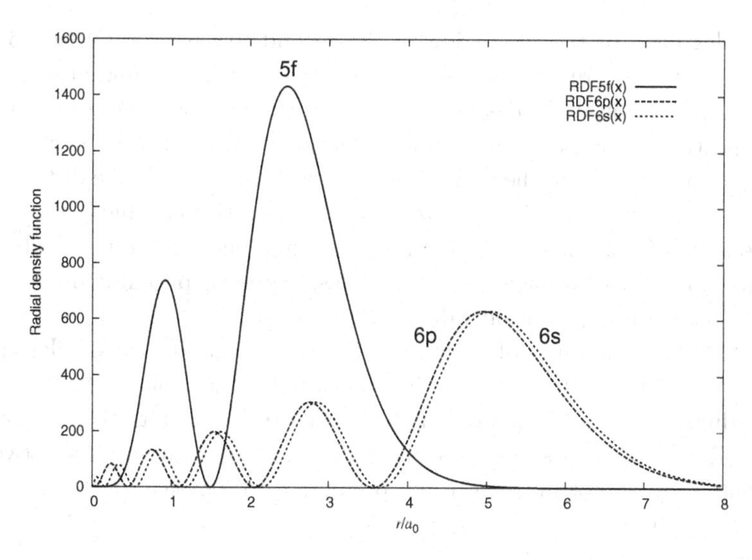

Figure 3.19. Radial density functions for the 5f, 6s and 6p subshells in U^{3+} using Slater's rules.

shows the analogous radial density functions for the 5f, 6s and 6p orbitals in trivalent uranium, U^{3+}.[4] We write the electron configuration of the ion has $[_{36}\mathrm{Kr}]\ 4d^{10}\ 4f^{14}\ (5s\ 5p)^8\ 5d^{10}\ 5f^3\ (6s\ 6p)^8$. The effective nuclear charges are

$$Z_{\text{eff}}(5\text{f}) = 92 - [(0.35 \times 2)(1.00 \times 78)] = 13.30 \qquad (3.13\text{a})$$

$$Z_{\text{eff}}(6\text{sp}) = 92 - [(0.35 \times 7)(0.85 \times 21)(1.00 \times 60)] = 11.70, \qquad (3.13\text{b})$$

which are the same as was just found for Nd^{3+}.

The expectation values, $\langle r \rangle$, for the electrons in the orbitals in Figures 3.18 and 3.19 are shown in Table 3.13 using the Z_{eff} values with equation (2.42).

The numbers in the table appear consistent with the plots of the radial density functions in s 3.18 and 3.19. The ratios of the weighted-average s and p expectation values to the f expectation value are 2.32 for the lanthanide and 1.92 for the actinide. This is consistent with the f orbitals not being so markedly core-like in the actinides as in the lanthanides. Energetic factors will of course be crucial in accounting for the different chemistry of the lanthanides and actinides.

[4]Figure 3.19 neglects the spin–orbit splitting of the 5f and 6p atomic orbitals.

Analogously to the discussion in the second-row part of Section 3.10, the decreasingly core-like nature of the outer f orbitals compared with the outer s and p orbitals on descending the f block might have been predicted by inspection of the expectation value equation (2.42). The argument about the larger values of n in the d orbitals making the most probable distance of the electron from the nucleus greater than the expectation value (on account of the increasing number of radial nodes) also applies to the f orbitals; for the 4f orbitals, which have no radial nodes, the most probable distance is smaller than the expectation value.

The published values of ionic radii for these ions [71] are smaller than would be expected from the radial density function plots: 1.86 a_0 for 6-coordinate Nd^{3+} and 1.94 a_0 for 6-coordinate U^{3+}. It has been reported [38] that while Slater's rules are useful for making comparisons between elements they do not make accurate predictions of physical quantities.

Chapter 4

Wavefunctions in Multi-Electron Atoms

4.1 Antisymmetric Wavefunctions: Slater Determinants

A fundamental property of fermions, and therefore electrons, is that wavefunctions describing more than one particle must change sign when any two particles are exchanged. Ultimately fundamental particles that are part of the same quantum system are indistinguishable and the order in which we number them cannot affect any physical property, i.e. $(\psi_a(1)\psi_b(2))^2 = (\psi_b(1)\psi_a(2))^2$. In squaring the wavefunctions rather than multiplying by their complex conjugate, they are assumed to be real, but that doesn't affect the conclusion. Taking the square root of the equation gives:

$$\psi_a(1)\psi_b(2) = \pm\psi_a(2)\psi_b(1). \tag{4.1}$$

The positive root pertains for bosons (fundamental particles with integral intrinsic spin, such as photons) and the negative root, imparting antisymmetry on the exchange of two particles, pertains for fermions. When $\psi_a = \psi_b$ it follows that

$$\psi_a(1)\psi_a(2) = -\psi_a(1)\psi_a(2) \tag{4.2}$$

since the wavefunctions commute. This expression can only be satisfied when $\psi_a(1)\psi_a(2) = 0$. This is consistent with the Pauli principle, as it is tantamount to that state being forbidden. In 1940 Pauli postulated that all fundamental particles are described by wavefunctions which are either symmetric (bosons with integer or zero spin) or antisymmetric (fermions with half-integer spin) with respect to the interchange of the full coordinates (position and spin) of a pair of identical particles. He also postulated that particles can never go from one symmetry type to another.

The ground state of helium is $1s^2$ with the two electrons' spins opposed. Even if we correct the 1s wavefunction to take into account the extra proton in the nucleus and the shielding from the other electron, we cannot write the wavefunction as $\Psi = 1s(1)\alpha(1)1s(2)\beta(2)$ because Ψ does not change sign when the electron labels (1) and (2) are exchanged. Capital Ψ denotes a multi-electron wavefunction. 1s(1) is shorthand for $\psi(n = 1, l = 0, m_l = 0)$ for electron (1) and $\alpha(1)$ is short for $\psi\left(s = \frac{1}{2}, m_s = \frac{1}{2}\right)$, for electron (1), etc.

Since the two electrons in $1s^2$ have the same orbital wavefunction, the multi-electron wavefunction is made antisymmetric with the spin wavefunctions:

$$\Psi_{1s^2} = \frac{1}{\sqrt{2}}\left(\psi_{1s}(1)\alpha(1)\psi_{1s}(2)\beta(2) - \psi_{1s}(2)\alpha(2)\psi_{1s}(1)\beta(1)\right), \tag{4.3}$$

where $\frac{1}{\sqrt{2}}$ is there to normalise the wavefunction. On squaring the function, this becomes $\frac{1}{2}$. Since the two terms in the wavefunction are already normalised and are orthogonal the cross terms vanish when the wavefunction is squared, leaving the wavefunction properly normalised. Inspection of equation (4.3) shows that exchange of the electron labels (1) and (2) reverses the sign of the wavefunction, as required.

It is instructive to consider the first excited configuration of helium, $1s^1 2s^1$. If both electrons have the same spin function then exchanging the electrons will leave the spin function unchanged. In this case, the orbital wavefunctions are made antisymmetric instead:

$$\Psi_{1s^1 2s^1} = \frac{1}{\sqrt{2}}\left(\psi_{1s}(1)\alpha(1)\psi_{2s}(2)\alpha(2) - \psi_{1s}(2)\alpha(2)\psi_{2s}(1)\alpha(1)\right), \tag{4.4}$$

and a similar wavefunction when both electrons' spins are β. A more convenient notation for these wavefunctions is to use determinants. When equation (4.4) is factorised and expressed in determinantal form it becomes

$$\Psi_{1s^1 2s^1} = \frac{1}{\sqrt{2}}\alpha(1)\alpha(2)\begin{vmatrix} \psi_{1s}(1) & \psi_{1s}(2) \\ \psi_{2s}(1) & \psi_{2s}(2) \end{vmatrix}. \tag{4.5}$$

A more compact notation for these determinants is just to give the diagonal elements, making equation (4.4) look like

$$\Psi_{1s^1 2s^1} = \frac{1}{\sqrt{2}}\alpha(1)\alpha(2) \mid \psi_{1s}(1) \quad \psi_{2s}(2) \mid. \tag{4.6}$$

When one electron's spin is α and the other is β, further combinations are required to make the spin wavefunction unambiguously symmetric or antisymmetric. The symmetric combination, in compact determinantal form, is

$$\Psi_{1s^12s^1} = \frac{1}{\sqrt{2}} \mid \psi_{1s}(1) \;\; \psi_{2s}(2) \mid \frac{1}{\sqrt{2}} (\alpha(1)\beta(2) + \alpha(2)\beta(1)) \,. \tag{4.7}$$

which, on expansion, is equivalent to

$$\Psi_{1s^12s^1} = \frac{1}{2} (\mid \psi_{1s}(1)\alpha(1) \;\; \psi_{2s}(2)\beta(2) \mid + \mid \psi_{1s}(1)\beta(1) \;\; \psi_{2s}(2)\alpha(2) \mid) \,. \tag{4.8}$$

The normalisation constant is $\frac{1}{2}$ so that when it is squared, $\frac{1}{4}$ can be in front of each of the four orthogonal and normalised terms so that the multi-electron function is properly normalised. The antisymmetric combination of the α and β spins is

$$\Psi_{1s^12s^1} = \frac{1}{\sqrt{2}} (\psi_{1s}(1)\psi_{2s}(2) + \psi_{1s}(2)\psi_{2s}(1)) \frac{1}{\sqrt{2}} \mid \alpha(1) \;\; \beta(2) \mid . \tag{4.9}$$

which, on expansion, is equivalent to

$$\Psi_{1s^12s^1} = \frac{1}{2} (\mid \psi_{1s}(1)\alpha(1) \;\; \psi_{2s}(2)\beta(2) \mid - \mid \psi_{1s}(1)\beta(1) \;\; \psi_{2s}(2)\alpha(2) \mid) \,. \tag{4.10}$$

In two-electron systems, either the orbital or the spin part of the overall wavefunction must be antisymmetric in order that the overall wavefunction is antisymmetric. If both parts were antisymmetric the overall wavefunction would be symmetric, rather like two minuses combining to give a plus.

The three spin-symmetric (and orbital anti-symmetric) states of He $1s^12s^1$ have the same energy and together form what is known as a triplet level. The remaining spin-antisymmetric (and orbital symmetric) state is known as a singlet state. Note that the word 'state' refers to a specified set of quantum numbers for every electron; 'level' can correspond to a set of states that have the same energy. Such states within the same level are known as degenerate. The number of states making up a degenerate level is known as the degeneracy of the level.

The formation of antisymmetric wavefunctions using determinants can be generalised to configurations of more than two electrons. An n-electron state can be described with an antisymmetric wavefunction by using an $n \times n$ determinant where the diagonal terms describe the orbital and spin state of each electron of the configuration in turn. For example the $1s^22s^1$ ground state of lithium with the 2s electron spin-up has the following antisymmetric wavefunction:[1]

$$\Psi_{1s^22s^1} = \frac{1}{\sqrt{3!}}\begin{vmatrix} \psi_{1s}(1)\alpha(1) & \psi_{1s}(2)\alpha(2) & \psi_{1s}(3)\alpha(3) \\ \psi_{1s}(1)\beta(1) & \psi_{1s}(2)\beta(2) & \psi_{1s}(3)\beta(3) \\ \psi_{2s}(1)\alpha(1) & \psi_{2s}(2)\alpha(2) & \psi_{2s}(3)\alpha(3) \end{vmatrix}. \tag{4.11}$$

An $n \times n$ determinantal wavefunction multiplies out to give $n!$ orthogonal terms, which accounts for the normalisation constant being $\frac{1}{\sqrt{n!}}$. These determinants were devised by Slater in 1929. Each entry in the determinant in known as a spin–orbital.

4.2 Energy Levels in Multi-Electron Atoms

To find the energy of a state we pre-multiply the Schrödinger equation of the antisymmetrised multi-electron wavefunction by the complex conjugate of that wavefunction. Since the wavefunction is normalised this gives the energy directly. We abbreviate the Schrödinger equation shown in equation (2.22) to $\hat{H}\Psi = E\Psi$. (Note that Z needs to be modified to a Z_{eff} to take into account the shielding of the nuclear charge by other electrons.) For a two-electron state there will need to be a second operator to act on the second electron, and also the $\hat{V}_{12}$ operator in equation (3.1) for the electrostatic potential energy of repulsion between the two electrons. In bra–ket notation, solving this equation for energy takes the form

$$\langle \Psi \mid H_1 + H_2 + V_{12} \mid \Psi \rangle = \langle \Psi \mid E \mid \Psi \rangle = E\langle \Psi | \Psi \rangle = E, \tag{4.12}$$

where H_1 is the combined kinetic and electron–nuclear potential energy of electron 1, etc. Since E is just a number it can be factorised out of the right-hand side of the equation.

The left-hand side of equation (4.12) can be factorised into three separate integrals, one for each of the operators. These too are often written in abbreviated forms, e.g. $\langle \Psi \mid H_1 \mid \Psi \rangle = \epsilon_1$. $\hat{H}_1$ and $\hat{H}_2$ are one-electron operators and so only operate on the relevant electron in the multi-electron wavefunction. The two-electron operator $\hat{V}_{12}$ is more complicated as it operates on two electrons at once. Let us consider the orbital

[1]This determinantal product wavefunction cannot be split up into an orbital-product sum and a spin-product sum as was done with the helium wavefunctions.

wavefunction of the two-electron configuration $1s^1 2s^1$. For convenience we use the additional shorthand:

$$u = 1s(1)2s(2) \quad \text{and} \quad v = 1s(2)2s(1).$$

When the two-electron orbital wavefunction is symmetric it is $\frac{1}{\sqrt{2}}(u+v)$ and when antisymmetric it is $\frac{1}{\sqrt{2}}(u-v)$. The energy operator acts in the position-space of x-, y- and z-coordinates, not the spin-space, and so for the purposes of this calculation we can ignore integrals involving spin. The only important consequence of the spin function is that it determines whether the orbital function is symmetric or antisymmetric (in order that the overall two-electron function is antisymmetric).

When the different operators act on the determinantal wavefunction, all possible combinations of u and v with the operator need to be considered. Some of these combinations will be equivalent, and others will come to zero because of the orthogonality of the atomic orbitals. Those that are not equal to zero are assigned a label as follows. Remembering that the one-electron operators only operate on the relevant electron we get the following results and labels:

$$\begin{aligned}
\langle u \mid H_1 \mid u \rangle = \langle v \mid H_2 \mid v \rangle &= \epsilon_{1s} \\
\langle u \mid H_2 \mid u \rangle = \langle v \mid H_1 \mid v \rangle &= \epsilon_{2s} \\
\langle u \mid V_{12} \mid u \rangle = \langle v \mid V_{12} \mid v \rangle &= J_{1s,2s} \\
\langle u \mid V_{12} \mid v \rangle = \langle v \mid V_{12} \mid u \rangle &= K_{1s,2s}.
\end{aligned} \tag{4.13}$$

The J integrals are known as direct-coulombic integrals; the K integrals are known as exchange-coulombic integrals and have no classical analogue. Both J and K integrals are positive. Other integrals, such as $\langle u \mid H_1 \mid v \rangle$, $\langle v \mid H_1 \mid u \rangle$, $\langle u \mid H_2 \mid v \rangle$, and $\langle v \mid H_2 \mid u \rangle$ are zero. Bearing in mind the factor $\frac{1}{2}$ from squaring the normalisation constant, the total energy of the $1s^1 2s^1$ state with a symmetric two-electron orbital wavefunction is therefore

$$E = \epsilon_{1s} + \epsilon_{2s} + J_{1s,2s} + K_{1s,2s}. \tag{4.14}$$

It is interesting to compare this result with the antisymmetric two-electron orbital wavefunction. The minus sign in front of v makes no difference when v is both the bra and the ket in the integral, so ϵ and J integrals are unaffected. However, the K integrals change sign. This means that the total energy of the $1s^1 2s^1$ state with an antisymmetric two-electron orbital wavefunction is

$$E = \epsilon_{1s} + \epsilon_{2s} + J_{1s,2s} - K_{1s,2s}. \tag{4.15}$$

The significance of this result is that triplet two-electron states, i.e. with symmetric spin functions and antisymmetric orbital functions, are lower in energy than singlet two-electron states, i.e. with antisymmetric spin functions and symmetric orbital functions.[2] This has far-reaching consequences in chemistry. The effect of the exchange integrals is to stabilise configurations where electrons are spinning in the same sense. In terms of the two-electron exchange integrals, there is an energetic advantage of $2K$ for two electrons to be spinning parallel rather than antiparallel.[3] This is the most important factor when considering the most stable state of an atom, known as the ground state. It should be emphasised that it is the position functions rather than the spin functions that determine the electron energy (in the non-relativistic picture, at least).

4.3 Russell–Saunders Terms

In order to consider the electronic spectroscopy of atoms, we need to be able to describe the overall electronic states of atoms. These states are the result of the coupling of angular momenta in atoms. The coupling scheme that best describes all but the heaviest atoms in the periodic table in the Russell–Saunders scheme, devised in 1925. This scheme produces so-called 'terms' described by two quantum numbers: the total spin, S, and the total orbital angular momentum, L. One way of arriving at the Russell–Saunders terms is to consider each possible arrow-in-box microstate in turn, i.e. considering individual electron angular momentum projection quantum numbers, m_s and m_l.

Exercise 45

For a subshell with orbital angular momentum quantum number l calculate the total number of possible microstates when there are n electrons in the subshell.

---

[2]This is discussed further in the context of the helium spectrum in Section 6.3.
[3]The direction of the spin angular momentum vector is assumed to point along the z-axis, but this is the projection ($\hbar m_s = \hbar/2$) of the overall spin ($\hbar\sqrt{s(s+1)} = \hbar\sqrt{3}/2$). It follows that the actual spin vector is at an angle of $\cos^{-1}(1/\sqrt{3}) = 54.7°$ to the z-axis (which is also the magic angle of NMR).

Exercise 46

Which f-subshell configuration is associated with the greatest number of possible microstates? What is the degeneracy of this configuration?

For each microstate the sum of the m_s of each electron gives the total spin projected on to the z-axis, M_S, and the sum of the m_l of each electron gives the total orbital angular momentum projected on to the z-axis, M_L. The capital-letter quantum numbers signify the total angular momentum from a multi-electron state. A convenient shorthand to describe microstates is to write the m_l value with a '+' superscript denoting $m_s = \frac{1}{2}$ (spin-up) and a '−' superscript denoting $m_s = -\frac{1}{2}$ (spin-down).

The microstates for the configuration p^2 are shown as an example in Table 4.1. We apply the principle that the electrons are indistinguishable (even though we label them 1 and 2 for convenience), and also the Pauli exclusion principle that no two electrons may be in an identical state. The number of possible microstates in p^2 is ${}^6C_2 = 15$. For each microstate the M_S and M_L value is given in the table.

Exercise 47

Write all the possible microstates of the subshell configuration p^3 in the shorthand style just described and label each microstate with its M_S and M_L value.

Table 4.1. The microstates of p^2 in the Russell–Saunders coupling scheme.

e(1)	e(2)	M_S	M_L
1^+	1^-	0	2
1^-	0^+	0	1
0^+	0^-	0	0
0^-	-1^+	0	−1
-1^+	-1^-	0	−2
1^+	0^+	1	1
1^+	-1^+	1	0
0^+	-1^+	1	−1
1^+	0^-	0	1
1^+	-1^-	0	0
0^+	-1^-	0	−1
1^-	0^-	−1	1
1^-	-1^-	−1	0
0^-	-1^-	−1	−1
1^-	-1^+	0	0

The M_S and M_L labels describing a complete set of microstates for an electron configuration can be arranged in groups described by the total angular momentum quantum numbers S and L. Just like the individual-electron quantum numbers, a term with an L total orbital angular momentum will be made up of microstates whose M_L labels span the range $-L$ to L in integer steps. The same is true for the range of M_S, meaning that each SL Russell–Saunders term is $(2S+1)(2L+1)$-fold degenerate. These SL groups are the terms that form the observable energy levels of atoms.

When there are two or more electrons in a subshell there will be microstates having the same values of M_S and M_L, and so there is a choice about the Russell–Saunders term to which to assign them. For the purpose of establishing which Russell–Saunders terms make up a configuration it doesn't matter which microstate is assigned to which term. In fact, suitably normalised linear combinations of these similar microstates are required to describe the eigenfunctions of Russell–Saunders terms containing these microstates.

The spectroscopic notation used to label each term is ${}^{2S+1}L$. The $(2S+1)$ is the spin multiplicity or degeneracy, i.e. the numbers of different possible projections of the total spin vector on to a quantised axis. When $2S+1 = 1$, i.e. when $S = 0$, the term is referred to as a singlet and when $2S+1 = 3$, i.e. when $S = 1$, the term is referred to as a triplet, etc. (this is the sense in which these words were used in Section 4.1). Instead of representing L with a number a letter is used. The convention is to follow the letters used for labelling subshells, i.e. S for $L = 0$, P for $L = 1$, D for $L = 2$ and F for $L = 3$. For successive values of L, successive letters after F, in alphabetical order, are used, except J is avoided due to its widespread use for total angular momentum. So when $L = 7$ the state is described as a K state.

The assignment of spectroscopic terms from the collection of electron microstates is demonstrated for the configuration p^2. Referring to Table 4.1, it is convenient to begin with the microstate with maximum M_L, which is 2. In order to maximise M_L, two electrons with maximum m_l must be spin-paired and so M_S will be equal to zero. This must belong to a Russell–Saunders term with $L = 2$ and $S = 0$ since the maximum M_L from a multiplet equals L. The other microstates that represent the degenerate projections of the L angular momentum will have M_L values decreasing in integer steps until $-L$ is reached. These are the four microstates under the $M_L = 2$ microstate in the table. These five microstates may be taken together to represent the Russell–Saunders term ${}^1\mathrm{D}$. The next highest M_L value is 1, which can be achieved with spin-parallel electrons so $M_S = 1$

may be associated with it. This belongs to a Russell–Saunders term with $S = 1$ and $L = 1$. Each of the three possible projections of M_L will be found with each of the three possible projections of M_S, giving a total degeneracy of 9. These nine microstates may be taken to be the ones beneath the top five that were associated with the ^{1}D state. These nine microstates make up the Russell–Saunders term ^{3}P. The only remaining microstate has $M_L = 0$ and $M_S = 0$. This is assigned to the Russell–Saunders term with $S = 0$ and $L = 0$, i.e. ^{1}S.

Exercise 48

Work out all the Russell–Saunders terms in the subshell configuration p^3 and label each term using spectroscopic notation.

Exercise 49

Which Russell–Saunders terms in the f block have the largest value of L?

Exercise 50

Work out all the Russell–Saunders terms in the subshell configuration pp (for example $2p^1 3p^1$) and label each term using spectroscopic notation. Explain why there are more terms than in the p^2 configuration.

Exercise 51

Considering the terms in pp and p^2, deduce the terms in dd, d^2, ff and f^2.

Exercise 52

Deduce the terms in p^1, p^4, p^5 and p^6.

In the Russell–Saunders coupling scheme, the only 'good' quantum numbers are S and L and their projections on the z-axis because the operators corresponding to these quantities commute with the Hamiltonian while other angular momentum operators do not. This is to say that S and L and their projections on the z-axis, which describe all the electrons on the atom as a whole, are the only quantum numbers that can be used directly to describe observable phenomena such as spectroscopic transitions and magnetic phenomena. Note that when there are more than two electrons in a d or f subshell not all the Russell–Saunders terms are unique. This introduces complications when considering the energies of the different terms. All the Russell–Saunders terms for all configurations in s, p, d and f subshells are collected in [72].

4.4 Spin $\frac{1}{2}$

The experimentally observed spin $\frac{1}{2}$ of the electron throws up some curious factor 2s. One of these is that two complete rotations (4π or $720°$) are required to return the electron to its original state. This is because quantum mechanical spin, which has no classical analogue, has a two-dimensional complex amplitude. This amplitude is transformed by the 2×2 complex Pauli matrices. The square of the complex matrices, corresponding to one complete rotation, returns -1 times the identity matrix, and so a further complete rotation is required to produce the identity matrix.

The spin $\frac{1}{2}$ nature of the electron also leads to a factor 2 with the magnetic moment. Treating the electron spin as if it were orbital angular momentum, i.e. as if it had integer angular momentum, a classical derivation considers an electron making circular orbits of the nucleus. In this model the magnetic moment on a given axis, μ_z, is the product of the electric current and the area enclosed by the current loop, πr^2. The current is the rate at which charge passes a fixed point, which is the charge on the electron, $-e$, multiplied by the frequency of its rotation in Hz. The frequency of the rotation in Hz is the linear velocity, v, of the electron divided by the distance of one complete rotation, $2\pi r$:

$$\mu_z = -e \times \frac{v}{2\pi r} \times \pi r^2 = -\frac{1}{2} evr. \tag{4.16}$$

This magnetic moment is typically written in terms of the classical angular momentum about the z-axis, $l_z = m_e vr$, where m_e is the mass of the electron:

$$\mu_z = -\frac{e}{2m_e} l_z. \tag{4.17}$$

The constant of proportionality, $-e/2m_e$, relating the angular momentum and the magnetic moment, is known as the magnetogyric ratio. While this derivation correctly predicts the magnetogyric ratio for orbital angular momentum, the ratio of the magnetic moment to the half-integral spin angular momentum is larger than predicted by a factor of 2. This factor 2 is known as the 'g-factor'. (The experimentally determined value of the g-factor is a fraction over 2.002; this is because the quantised electromagnetic vacuum contributes what is effectively a tiny zero-point energy [10].)

Another unexpected factor 2 arises when calculating the spin–orbit interaction (see Section 4.5). In this case the classically derived energy needs to be divided by 2 to obtain agreement with experiment. This factor $\frac{1}{2}$ is known as the Thomas precession.

4.5 Spin–Orbit Coupling

Until this point spin and orbital angular momenta have been considered independently of one another. It is however possible for them to interact magnetically since both angular momenta lead to a magnetic dipole (see Section 4.4), and magnets interact with one another. This energy of this interaction is related to Z^4, where Z is the atomic number [10], and is only significant for heavy atoms. For most atoms, spin–orbit coupling is referred to as 'fine structure'. The spin, $\mathbf{S}$, and orbital, $\mathbf{L}$, angular momentum vectors therefore couple to a resultant total angular momentum vector, $\mathbf{J}$:

$$\mathbf{S} + \mathbf{L} = \mathbf{J}. \tag{4.18}$$

The observed angular momenta projected on to a given axis of these three types are related to quantum numbers as was seen for the orbital angular momentum of an electron in hydrogen in equation (2.6):

$$L_z = \hbar M_L \tag{4.19a}$$

$$S_z = \hbar M_S \tag{4.19b}$$

$$J_z = \hbar M_J. \tag{4.19c}$$

It follows from equation (4.18) that the range of possible values of the J quantum number arising from the spin–orbit coupling of a Russell–Saunders term characterised by S and L quantum numbers is

$$|S-L|, \quad |S-L|+1, \ldots, \quad |S+L|-1, \quad |S+L|. \tag{4.20}$$

Following the arguments about the angular momentum of the hydrogen electron in Section 2.3, each spin–orbit coupled SLJ level is $(2J+1)$-fold degenerate. The spectroscopic notation used to label each spin–orbit level is ${}^{2S+1}L_J$.

Exercise 53

Using spectroscopic notation write down all the spin–orbit coupled levels of the p^2 configuration.

Table 4.2. The microstates of the ^{3}P term of p^2, showing their M_J values.

e(1)	e(2)	M_S	M_L	M_J
1^+	0^+	1	1	2
1^+	-1^+	1	0	1
0^+	-1^+	1	-1	0
0^+	-1^-	0	-1	-1
0^-	-1^-	-1	-1	-2
1^+	0^-	0	1	1
1^+	-1^-	0	0	0
1^-	-1^-	-1	0	-1
1^-	0^-	-1	1	0

Exercise 54

Work out the degeneracy of each spin–orbit coupled level of ^{3}P and show that the sum of these degeneracies is equal to the degeneracy of the ^{3}P term.

Spin–orbit coupling can also be applied to individual microstates, whose M_S and M_L values may be summed to give an M_J value. These M_J labels may be used to assign microstates to spin–orbit coupled J levels. This is demonstrated for the ^{3}P term of p^2 in Table 4.2.

Table 4.2 groups the microstates by their M_J values to make it clear how they might be assigned to ^{3}P$_2$ (the first five microstates), ^{3}P$_1$ (the next three) and ^{3}P$_0$ (the final one). These groupings are consistent with the $(2J+1)$-fold degeneracy of the J levels.

Exercise 55

Write out the microstates of the ^{2}D term of p^3, showing their M_J values and how they might be assigned to the spin–orbit coupled levels.

4.6 Hund's Rules

With knowledge of the spectroscopically determined ground states of gas-phase atoms, Hund devised three rules that predicted the ground-state level for any configuration, shown in the box.

Hund's Rules

(1) Find the term with largest S.
(2) When there are multiple terms with largest S, choose the term with the largest L.
(3) If the open subshell is less than half-full then the lowest-energy J level is the smallest value; if it is more than half-full, then the highest J level is the ground state.

The first rule is a result of the exchange energy discussed in Section 4.2. The second rule can be rationalised in terms of electrons avoiding each other; this may be achieved by the electrons revolving around the atom in the same direction so their paths don't cross. With the electrons revolving in the same directions their angular momenta will add to give a large resultant, maximising L. This rule accounts for the quarter- and three-quarter-filled subshell effect seen in the f block described in Section 3.11.

The third rule follows from the spin–orbit energy (see Chapter 6). The reason that the rule reverses when the subshell is more than half-full is that the spin–orbit interaction being considered is a one-electron interaction, i.e. the spin and orbital angular momenta being coupled are from the same electron. When a subshell is more than half-full it behaves as a spectroscopic 'mirror image' of the less-than-half-full subshell. In the 'mirror image' configurations positive holes take the place of electrons. Being positive, the associated spin magnetic moment reverses direction and the spin–orbit interaction changes sign. This reversal inverts the order of energies observed for the J levels within Russell–Saunders terms.

Exercise 56

Work out the ground Russell–Saunders term for each d configuration from d^1 to d^{10}. Comment on the symmetry of the answers.

Exercise 57

Work out the ground spin–orbit coupled level for each f configuration from f^1 to f^{14}.

Exercise 58

In the case of gas-phase nickel atoms the first excited state, at 205 cm^{-1} above the ground state, belongs to the $^3D_3$ level of the $4s^13d^9$ configuration. Use the Maxwell–Boltzmann expression in equation (4.21) to work out to the nearest percentage the proportion of atoms in a gas-phase sample of nickel atoms at 298 K that is in the $^3F_4$ level of the ground-state $4s^23d^8$ configuration. Neglect all the other excited states. Take 1 cm^{-1} to be equivalent to 1.986×10^{-23} J.

$$\frac{n_1}{n_0} = \frac{g_1 \exp\left(\frac{-E}{kT}\right)}{g_0}, \tag{4.21}$$

where $\frac{n_1}{n_0}$ is the ratio of the populations of excited state to ground-state atoms, g_1 and g_0 are the degeneracies of the excited and ground-state levels, respectively, E is the energy of the excited state above the ground state, k is the Boltzmann constant, 1.381×10^{-23} J K^{-1}, and T is the thermodynamic temperature.

4.7 *jj* Coupling

Spin–orbit coupling is normally considered to be a small perturbation to be added to the Russell–Saunders terms. The interaction becomes more important as atomic number increases, and is very important for lead and heavier elements. The *jj* coupling scheme assumes pure spin–orbit coupling producing spectroscopic states not depending on S and L quantum numbers at all. In practice, all spectroscopic ground states will depend on S and L, though some excited state configurations where the open-subshell electrons are widely separated exhibit virtually pure *jj* coupling [73]. This strong spin–orbit coupling regime is required for the description of states of protons and neutrons in the nucleus.

We saw in Section 4.3 that the six spin–orbitals of p^1 may be written as $1^+, 1^-, 0^+, 0^-, -1^+, -1^-$. To obtain the *jj* coupled states we first add the m_s and m_l quantum numbers to obtain an m_j quantum number to describe each state:

$$m_j = \frac{3}{2}, \frac{1}{2}, \frac{1}{2}, -\frac{1}{2}, -\frac{1}{2}, -\frac{3}{2}, \tag{4.22}$$

respectively. Inspection of the m_j values reveals that these values belong to $j = \frac{3}{2}$ and $j = \frac{1}{2}$ levels. Since $m_j = \frac{1}{2}$ and $m_j = -\frac{1}{2}$ appear twice then, strictly speaking, normalised symmetric and antisymmetric linear combinations of them need to be used. Considering the $(2j+1)$-degeneracy of these levels, it is clear that the number of microstates for p^1 has been preserved.

We saw in Section 4.3 that the p^2 configuration is made up of 15 microstates. p^2 in carbon is described well by the Russell–Saunders scheme, while p^2 in lead is better described by jj coupling. Following the procedure taken with p^1 we construct the $|j, m\rangle$ states for each electron and then add the m for each electron to give M for the two-particle state. It is important that the $|j, m\rangle$ state assigned to each electron is not the same as this would contravene the Pauli principle. From these M values the J values should become evident since each J level should be composed of M states running from $-J$ to J in integer steps, in common with other angular momenta. Using the $|j, m\rangle$ states from p^1 and neglecting the linear combinations for repeated m and M values we obtain the results in Table 4.3.

Note the systematic way in which the table is completed. In the first row is the microstate with the smallest j_1 and j_2, which may only be singly degenerate to preserve the Pauli principle. In the second row j_2 is incremented and m_2 takes its minimum value. Over subsequent rows m_2 is incremented until it reaches its maximum value. Then m_1 in incremented with m_2 taking its minimum value. Again m_2 is incremented in subsequent

Table 4.3. The microstates of p^2 in the jj coupling scheme.

j_1	m_1	j_2	m_2	M
1/2	−1/2	1/2	1/2	0
1/2	−1/2	3/2	−3/2	−2
			−1/2	−1
			1/2	0
			3/2	1
1/2	1/2	3/2	−3/2	−1
			−1/2	0
			1/2	1
			3/2	2
3/2	−3/2	3/2	−1/2	−2
			1/2	−1
			3/2	0
3/2	−1/2	3/2	1/2	0
			3/2	1
3/2	1/2	3/2	3/2	2

rows until it reaches its maximum value. Once the maximum value of m_1 has been exhausted, j_1 is incremented with m_1 and m_2 on their minimum values. m_2 and then m_1 are then incremented in a similar way to before.

From Table 4.3 it is evident that when $j_1 = \frac{1}{2}$ the resulting M values belong to levels with J equal to 0, 1 and 2. Similarly, when $j_1 = \frac{3}{2}$ the M values belong to levels with J equal to 0 and 2. A useful nomenclature for these states would be $|(j_1, j_2, \ldots)JM\rangle$. In practice the states of jj configurations are usually summarised as $(j_1, j_2, \ldots)_J$.

It is instructive to see which microstates appear to be missing from Table 4.3. For example the state $|(\frac{3}{2}, \frac{1}{2})2 \ -2\rangle$, whose two electrons are not equivalent, is not seen. This is because the state with the electrons ordered the other way round has been included. Due to the Pauli principle, it should strictly be the antisymmetric linear combination of these two states that is used. The symmetric combination is forbidden.

Exercise 59

Work out the 20 microstates of p^3 in jj coupling ignoring the necessary linear combinations. Name the levels using the $(j_1, j_2, j_3)_J$ system of nomenclature.

4.8 Other Types of Magnetic Coupling

In Section 4.5 on spin–orbit coupling we considered the the fine-structure interaction: magnetic coupling of the total spin angular momentum of the electrons with the total orbital angular momentum; in Section 4.7 on jj coupling we considered the magnetic coupling of the spin and orbital angular momenta on individual electrons before summing the contributions from the different electrons. Other types of magnetic coupling are possible, though the energies of their interaction are of smaller magnitude.

One such interaction is the hyperfine structure interaction. This involves the coupling of the magnetic field from the electron angular momenta with a nuclear magnetic moment associated with the nuclear spin quantum number, I. The coupling of these two angular momenta results in an overall angular momentum, F. The hyperfine interaction can be observed in hydrogen since the nucleus has spin $I = \frac{1}{2}$. The hydrogen electron in the 1s ground state has no orbital angular momentum but has its spin quantum number $s = \frac{1}{2}$. The s and I couple to give two energy levels, characterised by overall angular momenta of $F = 0$ and 1, split by about 0.0474 cm^{-1}, with $F = 0$ being the lower level [31].

Exercise 60

The separation of ^{235}U ($I = \frac{7}{2}$, 0.7% abundance) and ^{238}U ($I = 0$, 99.3% abundance) isotopes can, in principle, be achieved by making use of the hyperfine interaction. Suggest how this may be achieved.

In principle other magnetic interactions are possible but they are often small enough to be ignored. For example, the spin angular momentum on one electron may couple magnetically with the spin or the orbital angular momentum on another electron, known as the spin-other-spin and spin-other-orbit interactions, respectively.

4.9 Crystal Fields

While this book is principally concerned with atomic rather than condensed phase spectra, there are some interesting parallels between the two. Crystal fields will be considered in three contexts: their effect on a single electron, a weak field acting on multi-electron configurations, and a strong field acting on multi-electron configurations.

4.9.1 *Single-electron states*

Transition metal ions in octahedral complexes experience ligand interactions along the Cartesian axes that define the six ligand positions: x, $-x$, y, $-y$, z and $-z$. Hence orbitals within a subshell are differentiated according to the extent to which they point along Cartesian axes.

The three p orbitals are symmetrically equivalent with respect to the Cartesian axes, each pointing directly along an axis, and so are not split by the ligands in an octahedral complex; they are given the symmetry label T_1 in octahedral symmetry. The s orbital isn't affected either and takes the fully symmetric A_1 label in octahedral symmetry.

The five d orbitals (shown in Figure 2.4) either have lobes pointing between a pair of Cartesian axes (d_{xy}, d_{xz} and d_{yz}) or directly along Cartesian axes (d_{z^2} and $d_{x^2-y^2}$). This is apparent from the label on each orbital: products of different axes indicating lobes pointing between these axes, and powers of an axis indicating lobes pointing along the axis.

Inspection of the f orbitals in cubic symmetry (shown in Figure 2.7) reveals an orbital pointing between all three Cartesian axes (f_{xyz}), three

Table 4.4. Symmetry labels applied to octahedral and tetrahedral complexes and their degeneracy, g. The 180° rotation refers to one around an axis perpendicular to the principal axis.

Label	g	Distinguishing property
A_1	1	Phases unchanged by 180° rotation
A_2	1	Phases reversed by 180° rotation
E	2	
T_1	3	Phases reversed by 180° rotation
T_2	3	Phases unchanged by 180° rotation

Table 4.5. The cubic crystal field levels applied to the subshells and the orbitals that take the labels.

Subshell	Cubic label	Orbitals
s	A_1	s
p	T_1	p_x, p_y, p_z
d	E	d_{z^2}, $d_{x^2-y^2}$
	T_2	d_{xy}, d_{xz}, d_{yz}
f	A_2	f_{xyz}
	T_1	f_{x^3}, f_{y^3}, f_{z^3}
	T_2	$f_{x(z^2-y^2)}$, $f_{y(z^2-x^2)}$, $f_{z(x^2-y^2)}$
g	A_1	$g_{x^4+y^4+z^4}$
	E	$g_{z^2(x^2-y^2)}$, $g_{2z^4-x^4-y^4}$
	T_1	$g_{xz(z^2-x^2)}$, $g_{yz(z^2-y^2)}$, $g_{xy(x^2-y^2)}$
	T_2	g_{xyz^2}, g_{xzy^2}, g_{yzx^2}

symmetrically equivalent orbitals each pointing directly along an axis (f_{x^3}, f_{y^3} and f_{z^3}) and three other symmetrically equivalent orbitals whose labels involve a cyclic permutation of the axes ($f_{x(z^2-y^2)}$, $f_{y(z^2-x^2)}$ and $f_{z(x^2-y^2)}$).

Symmetry labels are assigned to the groupings of symmetrically equivalent orbitals in octahedral symmetry. They are given in Table 4.4 with their degeneracy, g, and a property that distinguishes labels of the same degeneracy. A summary of the octahedral crystal-field labels that apply to each subshell is given in Table 4.5.

The greater the extent to which an orbital's lobes point at the ligands the higher that orbital's energy relative to orbitals in the same subshell pointing between the axes. In a dipolar model this is taken to reflect electrostatic repulsion between the electrons on the transition metal ion and the ligand electrons. In a molecular orbital treatment (see Section 5.1) of transition metal ions in octahedral complexes, the T_2 orbitals, d_{xy}, d_{xz} and d_{yz}, pointing between the ligands are non-bonding while the E orbitals, d_{z^2} and $d_{x^2-y^2}$, pointing at the ligands are higher in energy because they are anti-bonding (the ligand lone pairs are in the bonding orbitals). It follows then that of the f orbitals the A_2 f_{xyz} ought to be lowest in energy in an octahedral complex while the T_1 orbitals, f_{x^3}, f_{y^3} and f_{z^3}, ought to be highest in energy.

In tetrahedral complexes the orbitals may be given the same basic labels as in octahedral complexes as both symmetries are in the cubic family. The centres of the six faces of the cube make up the six corners of an octahedron; indeed this process also applies in reverse, making the two shapes duals. Taking half of the eight corners of a cube, each of them diagonally opposite the other three on the shared faces, gives the corners of a tetrahedron. The geometric relationships of the octahedral and tetrahedral complexes to a cube are shown in Figure 4.1. Since the ligands are at corners of a cube in tetrahedral complexes (as opposed to being at the centre of the cube's faces in octahedral complexes) the ordering of the orbitals' energies is reversed compared with octahedral complexes.

The important difference between the symmetry of a tetrahedron and an octahedron is that only the latter has a centre of symmetry. For this reason, in octahedral symmetry, a subscript g or u is given to symmetry labels where g denotes even symmetry with respect to inversion, and u odd symmetry.

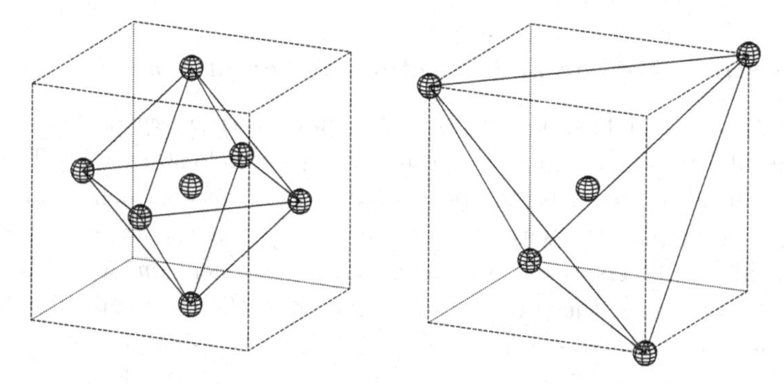

Figure 4.1. The geometric relationship with a cube of octahedral and tetrahedral complexes.

4.9.2 *Weak field on multi-electron configurations*

When a weak crystal field acts on a Russell–Saunders term associated with a multi-electron configuration, the term splits according to the quantum number L in the same way that a single-electron state splits according to its subshell l. The spin quantum number S is unaffected. For example, the ^{3}F (degeneracy $g = 21$) ground state of d^2 in an octahedral complex will split into $^3A_{2g}(g = 3)$, $^3T_{2g}(g = 9)$ and $^3T_{1g}(g = 9)$ levels. The g values indicate that the total number of states of the level is unaffected by the crystal field interaction.

Exercise 61

Suggest how the ^{1}G term of d^2 in an octahedral complex will split. Use the degeneracies to show that the total number of states remains unchanged.

In the lanthanides the 4f orbitals are quite core-like, given their high angular momentum and the absence of any radial nodes. The interaction of lanthanide 4f electrons with ligands is weak compared with the spin–orbit interaction. Hence the crystal field acts on the $^{2S+1}L_J$ levels. The J quantum number is split by the field in the same way as the L quantum number. The splitting of levels of L greater than 4 and of half-integral J levels is explained in [28].

Exercise 62

Into what crystal-field levels is the ground-state level of the Pr^{3+} configuration split when the free atom becomes octahedrally complexed?

4.9.3 *Strong field on multi-electron configurations*

When the crystal field is strong, individual electron crystal-field states couple with one another to give a resultant crystal-field state. Only T states will be considered here, as their degeneracy of 3 means they behave in a way that has some parallels with the p subshell [27]. The theory of E states is explained in [28]. The linear combinations of component $|m\rangle$ wavefunctions (from Table 2.2) of the p orbitals and the cubic T_2 d orbitals are shown in Table 4.6. There is a clear parallel with the p subshell for d_{xz} and d_{yz} orbitals, but less so with the d_{xy} orbital. Among the d orbitals, it is the cubic E orbital d_{z^2} that mirrors the p_z orbital.

Table 4.6. The p and cubic T_2 d orbitals expressed as linear combinations of their component $|m\rangle$ wavefunctions.

Orbital	Linear combination
p_x	$|1\rangle + |-1>$
p_y	$|1\rangle - |-1>$
p_z	$|0\rangle$
d_{xz}	$|1\rangle + |-1>$
d_{yz}	$|1\rangle - |-1>$
d_{xy}	$|2\rangle - |-2>$

We saw in Section 4.3 that two p electrons form ^{3}P, ^{1}D and ^{1}S terms. To find the strong-field states of two electrons in T octahedral crystal-field states, the Russell–Saunders terms of p^2 (degeneracy $g = 15$) are reduced to their octahedral crystal-field states following the methods discussed for weak crystal fields. This results in the crystal-field states $^3T_{1g}(g = 9)$, $^1E_g(g = 2)$, $^1T_{2g}(g = 3)$ and $^1A_{1g}(g = 1)$ which are the states that describe two electrons in T octahedral field states. Note that summing the degeneracies of all the crystal-field states gives 15, the degeneracy of p^2. This T×T result applies equally to electrons in T_1 and T_2 states.

Exercise 63

Work out all the octahedral crystal field states when there are four, five and six electrons in a T crystal-field level.

The analogy between T^n and p^n configurations doesn't quite extend to three electrons on account of the dissimilarity between d_{xy} and p_z. The p^3 model doesn't quite predict the actual T^3 octahedral crystal-field levels, $^4A_{2g}$, ^2E_g, $^2T_{2g}$ and $^2T_{1g}$.

Exercise 64

What crystal field states does the p^n model predict when there are three electrons in a T crystal-field level?

In first-row transition elements, the crystal field experienced by the d electrons is often intermediate between the weak- and strong-field schemes, making the interpretation of spectra more complicated [27].

Chapter 5

Orbitals in Molecules

The chemical bonds that hold together atomic nuclei in molecules are shared electrons between the nuclei. The most widely applied approach to chemical bonding in molecules is molecular orbital (MO) theory, though there has been increased interest recently in valence bond theory [74], the original approach taken to quantum chemical calculations by Heitler and London in 1927 [75]. MO theory is based upon wavefunctions constructed from the linear combination of the atomic orbitals that overlap to form the different bonding (or antibonding) interactions.

5.1 The Linear Combination of Atomic Orbitals (LCAO)

The overlap of atomic orbitals is normally first considered pictorially, with lobes of the same phase reinforcing each other to form a bonding molecular orbital and those of different phase cancelling, leaving an antibonding molecular orbital. It should be emphasised that the overlapping lobes are from the wavefunction itself, not the probability density; this is a reason that orbitals are principally taken to represent the wavefunction these days, rather than the probability density.

The bonding orbitals may be considered to be the result of constructive interference between standing-wave orbitals, causing an increase in amplitude, while antibonding orbitals result from destructive interference. This picture adds further weight to the idea of electrons as waves. In a bonding interaction the overlap of the orbitals extends the length of the orbital. This means that the resulting standing wave will have a longer wavelength, less curvature and therefore a lower kinetic energy. The opposite applies to antibonding interactions.

In the bonding molecular orbital most of the electron density is between the nuclei. It is usually taught that the electrostatic attraction between the nuclei and the bonding electrons, i.e. the electrostatic potential energy, constitutes the chemical bond. Detailed quantum chemical studies have shown that the lowering of the kinetic energy is the most important contribution and that the potential energy contribution can be repulsive [76]. In the antibonding orbital most of the electron density is on the other side of each nucleus, so that the electrostatic repulsion between the atoms is maximised. When such a state is occupied, it may result in the two atoms becoming completely disconnected.

In this pictorial approach to LCAO theory, the relative orientation of the phases of the atomic-orbital lobes determines whether interactions are bonding or antibonding. This is somewhat problematic for the overlap of two 1s orbitals, say, where the phase is only positive in both orbitals. However, following the properties of eigenfunctions (Section 1.4) we know that we can choose any value for the constant in front of the eigenfunction, with it remaining a valid solution to the eigenvalue (Schrödinger) equation. The sign is therefore unrestricted so we can assign phases to atomic orbital wavefunctions.[1] In this way, one of two overlapping atomic s orbitals can be given a negative phase arbitrarily, allowing for an antibonding combination. All of the possible phase combinations of atomic orbitals need to be considered when constructing a molecular orbital diagram.

Figure 5.1 shows a sketch of the molecular orbital diagram[2] for N_2. The relative energy levels of the valence atomic orbitals are shown on the right and left, with the molecular orbitals shown in order of energy up the middle. The dotted lines indicate which atomic orbitals combine to form a particular molecular orbital. The most important interactions are shown between atomic orbitals of the same type. Since it is a homonuclear diatomic, the like orbitals will have identical energy, which maximises the interaction between them. Antibonding orbitals are denoted with an asterisk. The bond order of the molecule is equal to the number of fully occupied bonding orbitals minus the number of fully occupied antibonding orbitals.

[1] The real consideration is the *relative* phases assigned to the s orbitals in the linear combination; two negative-phase s orbitals will give a bonding interaction.

[2] In reality the upper σ^* orbital in a molecular orbital diagram is often so high in energy that it 'dissolves' in the ionisation continuum and is not available for electronic occupation.

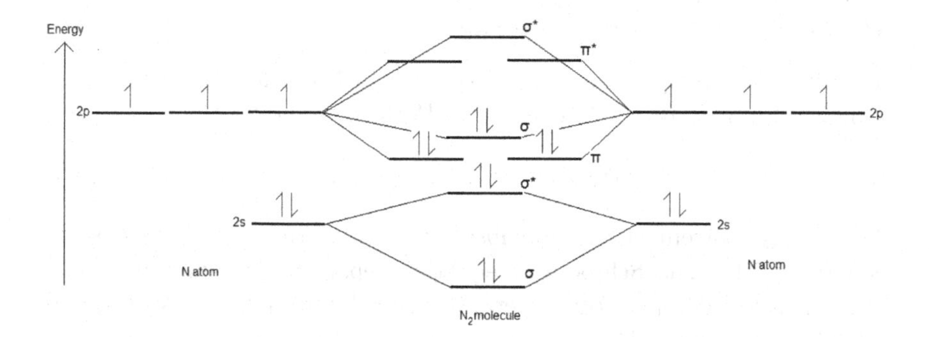

Figure 5.1. Sketch of the molecular orbital diagram for N_2.

As Figure 5.1 makes clear, each molecular orbital, like atomic orbitals, can only accommodate two electrons. This is, again, a consequence of the Pauli principle: each molecular orbital can be characterised by a number of quantum numbers, and so two electrons can fill it as long as their spins, i.e. their m_s quantum numbers, are different.

5.2 Molecular Orbital Energies

We rearrange the Schrödinger equation[3] and attempt to solve it using a linear combination of atomic orbitals, ψ_{MO}, in place of a single atomic orbital:

$$\langle \psi_{\mathrm{MO}} \hat{H} | \psi_{\mathrm{MO}} \rangle - E \langle \psi_{\mathrm{MO}} | \psi_{\mathrm{MO}} \rangle = 0 \tag{5.1}$$

$$\psi_{\mathrm{MO}} = \sum_i c_i \psi_i. \tag{5.2}$$

$\hat{H}$ is the Hamiltonian operator for total energy, i is a label for each atomic orbital overlapping to form the molecular orbital, and c_i is the coefficient in the summation for atomic orbital i. Our purpose is to find the different c_i to characterise the molecular orbital wavefunctions that solve equation (5.1), and to find their energies, E. The number of molecular orbitals that solve such a Schrödinger equation is equal to the number of atomic orbitals in the linear combination; indeed this approach conserves the number of orbitals before and after overlap. We will begin by applying MO theory to homonuclear and heteronuclear diatomic molecules.

[3]We use the equation in its simplest form, with an effective one-particle operator.

Homonuclear diatomic molecules

The simplest example is hydrogen, H_2, in which we take the overlapping atomic orbitals to be the 1s orbitals on H atoms 1 and 2:

$$\psi_{\mathrm{MO}} = c_1\psi_{1\mathrm{s}}(1) + c_2\psi_{1\mathrm{s}}(2). \tag{5.3}$$

This ψ_{MO} wavefunction is substituted into the ket positions in equation (5.1). The Schrödinger equation separates the sums into one-electron integrals. Each component atomic orbital of ψ_{MO} is substituted into the bra positions in turn, resulting in n separate equations, where n is the number of atomic orbitals overlapping to form the molecular orbitals. These n equations are solved simultaneously to obtain the values for c_i and E for the n molecular orbitals that solve the equation. In our H_2 example we need to solve two equations simultaneously. Since the integrals will not be explicitly solved, they are written in abbreviated form as follows:

$$\langle\psi(1)\hat{H}|\psi(2)\rangle = H_{12} \tag{5.4a}$$

$$\langle\psi(1)|\psi(2)\rangle = S_{12}. \tag{5.4b}$$

The equations to be solved simultaneously are usually written in this abbreviated notation.

$$\begin{aligned} c_1(H_{11} - ES_{11}) + c_2(H_{12} - ES_{12}) &= 0 \\ c_1(H_{21} - ES_{21}) + c_2(H_{22} - ES_{22}) &= 0 \end{aligned} \tag{5.5}$$

Equations (5.5) are known as secular equations (from the Latin *saeculum*, meaning 'once a century' since these equations were first applied to astronomical phenomena that repeated about once a century).

The H_{11} and H_{22} integrals are the energy of the 1s electron on H atoms 1 and 2, which will be identical, and are further abbreviated to α. The S_{11} and S_{22} integrals will, by definition, be equal to 1 when the atomic orbital wavefunctions are normalised. By symmetry $H_{12} = H_{21}$; these are both further abbreviated to β, known as resonance integrals. These β integrals turn out to be negative since the attractive electron–nuclear forces outweigh the repulsive inter-electron forces, and attractive forces lead to negative interaction energies. By symmetry $S_{12} = S_{21}$; these are both further abbreviated to S, known as overlap integrals. These overlap integrals are between 0 and 1, depending on the extent of the overlap. Applying these additional simplifications and abbreviations the secular equations take a simpler form:

$$c_1(\alpha - E) + c_2(\beta - ES) = 0$$
$$c_1(\beta - ES) + c_2(\alpha - E) = 0. \tag{5.6}$$

Equations (5.6) can be trivially solved by setting the c_i constants to zero (negating all wavefunctions). They can be more usefully solved by solving the $n \times n$ determinant of the constants attached to the $n\, c_i$ parameters:

$$\begin{vmatrix} \alpha - E & \beta - ES \\ \beta - ES & \alpha - E \end{vmatrix} = 0. \tag{5.7}$$

The 2×2 determinant has the general solution

$$\begin{vmatrix} a & b \\ c & d \end{vmatrix} = ad - bc. \tag{5.8}$$

We apply equation (5.8) to the determinant in equation (5.7):

$$(\alpha - E)^2 = (\beta - ES)^2$$
$$\alpha - E = \pm(\beta - ES)$$
$$E = \frac{\alpha \pm \beta}{1 \pm S}. \tag{5.9}$$

As expected, there are two solutions for energy: one is the bonding energy and the other the antibonding energy, with the bonding energy taking the lower value. Since β is negative we can deduce

$$E_{\mathrm{MO,bond}} = \frac{\alpha + \beta}{1 + S}$$
$$E_{\mathrm{MO,antibond}} = \frac{\alpha - \beta}{1 - S}. \tag{5.10}$$

The expressions for E in equations (5.10) can be inserted back into the secular equations (5.6) to reveal the relationships between c_1 and c_2. The outcome is as follows:

$$c_1 = c_2 \,(\text{bonding})$$
$$c_1 = -c_2 \,(\text{antibonding}). \tag{5.11}$$

Exercise 65

Show that the results of equation (5.11) are true.

The relative signs of the c_i coefficients reflect the in-phase and out-of-phase overlap of the hydrogen 1s orbitals in forming bonds and antibonds, respectively. Having established that c_1 and c_2 have the same magnitude in the MO wavefunctions, which isn't surprising considering the symmetry of the molecule, the value they take is determined by the normalisation of the wavefunction, leading to the following wavefunctions:

$$\psi_{\text{MO,bond}} = \frac{1}{\sqrt{2}}(\psi_{1s}(1) + \psi_{1s}(2))$$

$$\psi_{\text{MO,antibond}} = \frac{1}{\sqrt{2}}(\psi_{1s}(1) - \psi_{1s}(2)). \tag{5.12}$$

Exercise 66

Show that the value taken by c in equations (5.12) is true.

Since S must be between 0 and 1, equations (5.10) show that the antibonding MO is further above the non-bonding atomic energy α than the bonding MO is beneath it in energy. The antibond is therefore more antibonding than the bond is bonding. This explains why He_2 is not a stable molecule. Given that the MO diagram for He_2 predicts a bond order of 0, the molecule might have been assumed to be in equilibrium with free He atoms.

This excess antibonding idea can also be used to explain why the F$-$F bond energy is so weak. The bond order of 1 that is derived from its MO diagram is the result of subtracting two antibonds from three bonds. Given that there is scope for significant π overlap between the small fluorine atoms, the excess antibonding energy is likely to be significant, which should diminish somewhat the formal bond order of 1. Indeed, being the smallest atom in period 2 that forms bonds, and having all its π orbitals full, this bond-weakening effect should be more important in F_2 than in any other molecule.

Heteronuclear diatomic molecules

When two different types of atom are bonded together, we can no longer assume that $H_{11} = H_{22}$. They are abbreviated to α_1 and α_2, which approximate the orbital energies for those atoms. Atoms with higher ionisation energies have lower α values: the electron energy at the ionisation limit, where it is just free from the influence of the atom, may be taken to be a common energy for all atoms, often defined as $E = 0$.

With two different types of atom bonding there will be an orbital mismatch in the bonding interaction. It is therefore reasonable to assume, to

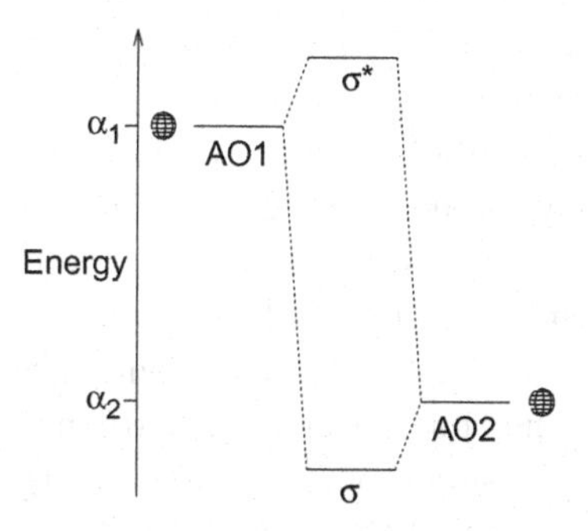

Figure 5.2. Sketch of the molecular orbital diagram for an idealised heteronuclear diatomic molecule.

a first approximation, that the overlap integral is 0, which helps counteract the added complication from having two α terms. For a heteronuclear diatomic molecule the secular determinant that solves the Schrödinger equation may be taken to be

$$\begin{vmatrix} \alpha_1 - E & \beta \\ \beta & \alpha_2 - E \end{vmatrix} = 0. \tag{5.13}$$

The approximate solutions to this equation are as follows:

$$\begin{aligned} E_1 &\approx \alpha_1 + \frac{\beta^2}{\alpha_1 - \alpha_2} \\ E_2 &\approx \alpha_2 - \frac{\beta^2}{\alpha_1 - \alpha_2}. \end{aligned} \tag{5.14}$$

When $\alpha_1 > \alpha_2$, E_1 is the antibonding energy and E_2 is the bonding energy. When $\alpha_2 > \alpha_1$, the energy assignments are reversed. A sketch of the molecular orbital diagram for an idealised heteronuclear diatomic molecule is shown in Figure 5.2, showing the α_1 atomic orbital energy higher than α_2, as we are supposing.

Exercise 67

Show how to solve equation (5.13). Hint: express the square rooted term in the solution to the quadratic equation as $\sqrt{1+x}$. If the x can be justified as being much smaller than 1 then $\sqrt{1+x} \approx 1 + \frac{1}{2}x$.

Taking E_1 as the energy of the antibonding molecular orbital, this orbital will more closely resemble the higher energy $\psi(1)$ atomic orbital (with energy α_1) than the lower energy $\psi(2)$ atomic orbital (with energy α_2); the latter atomic orbital more closely resembles the bonding molecular orbital (with energy E_2). In HF, for example, the more electronegative F atom will have valence electrons at lower energy than the H electron. So in our scheme, HF's $\psi(1)$ would belong to H and $\psi(2)$ to F, and the bonding MO would more closely resemble the fluorine AO.

The resemblance between atomic and molecular orbitals in energy terms is also reflected in terms of the coefficients c_1 and c_2 in the molecular orbital wavefunctions. As before, the coefficients for each molecular orbital are found by substituting its energy back into the secular equations. We substitute the E_1 expression from equation (5.14) into the first secular equation and rearrange to find the ratio c_1/c_2:

$$c_1\left(\alpha_1 - \alpha_1 - \frac{\beta^2}{\alpha_1 - \alpha_2}\right) + c_2\beta = 0$$

$$\frac{c_1}{c_2} = \frac{\alpha_1 - \alpha_2}{\beta}. \tag{5.15}$$

The numerator will be larger than the denominator in equation (5.15), so c_1 will be larger than c_2, consistent with the antibonding molecular orbital resembling the atomic orbital $\psi(1)$. c_1 and c_2 will also be of opposite sign since β is negative. This is consistent with the antibonding molecular orbital resulting from the atomic orbitals being out of phase with one another. Now we substitute the E_2 expression from equation (5.14) into the second secular equation to find c_1/c_2.

$$c_1\beta + c_2\left(\alpha_2 - \alpha_2 + \frac{\beta^2}{\alpha_1 - \alpha_2}\right) = 0$$

$$\frac{c_1}{c_2} = -\frac{\beta}{\alpha_1 - \alpha_2} \tag{5.16}$$

This result indicates that c_2 will be larger than c_1 for the bonding molecular orbital, consistent with it resembling the atomic orbital $\psi(2)$. With β being negative, c_1 and c_2 will be of the same sign, consistent with the bonding molecular orbital resulting from the overlap of atomic orbitals that are in phase with one another.

As we saw with homonuclear diatomics, the β term reflects the bonding interaction, while α reflects the atomic orbital energies. The energies of equation (5.14) indicate that the extent of the bonding interaction reduces as the atomic orbital energies diverge more widely. This reflects an increasing spatial mismatch between the orbitals that will accompany the energy mismatch. In the limit of diverging atomic orbital energies we find ionic bonding: the electron in the high energy AO on atom 1 transfers to the low energy AO on the other atom, and the attraction between the resulting positive and negative electric charges constitute the bond between the two atoms. At intermediate energy differences between the two atomic orbitals, the reduced covalent bonding interaction is compensated to an extent by the dipole in the bond that reflects the asymmetry in the bonding interaction, which adds an ionic component to the bond. This is a polar-covalent bond.

5.3 Types of Orbital Overlap

In Figure 5.1 we see that the 2p orbitals on each N atom overlap in two different ways, giving molecular orbitals carrying σ or π labels. These labels relate to the geometry of the overlap involved. Single chemical bonds always involve σ overlap, which is when the lobes of the two orbital are aligned head-on so that they meet directly between the two nuclei associated with the two orbitals. π overlap is when the orbitals overlap while they are sideways-on, i.e. the overlapping lobes are not pointing at each other. The regions of overlap therefore do not lie on the internuclear axis: they are between the nuclei but displaced from the internuclear axis. A maximum of two π bonds are found in multiple bonds, in addition to one σ bond (though there has recently been a suggestion that a quadruple bond forms in C_2 [74]).

Normally the head-on σ alignment allows for greater overlap between orbitals. This leads to a larger interaction and a greater energy gap between the resulting bonding and antibonding orbitals. One way of breaking a chemical bond is to promote an electron from the bonding orbital to the antibonding one. The more energy required to do this, the stronger the bond is considered to be. σ bonds are therefore generally stronger than π bonds. π bonds, however, are made stronger by delocalisation since extending the orbital lengthens the standing wave, increasing its wavelength and lowering its energy relative to its antibonding orbital.

In Figure 5.1 the energies of the molecular orbitals from the overlap of the 2p electrons do not seem consistent with this approach: given the absence of π delocalisation, one would expect the σ bonding orbital to be lower in energy than the π bonding one. The picture has been distorted by an effect known as s–p mixing. One way of rationalising it is that the two bonding σ orbitals have the same symmetry (even parity) and so are connected by the Hamiltonian operator, mixing their wavefunctions and causing them to diverge further in energy (as was seen with interacting energy levels in heteronuclear diatomics). Similarly the two antibonding σ^*

orbitals have the same symmetry (odd parity), interact and diverge further in energy. These interactions may also be visualised as resulting from the overlap of the s orbital on one atom with the p_z orbital on another. Alternatively, it may be considered as sp hybridisation (see Section 5.4).

s–p mixing is a smaller effect than the $2p_z$-$2p_z$ σ overlap since the 2s and $2p_z$ atomic orbitals are at different energies. Indeed the extent of the s–p mixing interaction increases as the energies of the two orbitals become more similar. This is the same effect as was seen with the atomic orbital energies in heteronuclear diatomics (see Section 5.2). Moving across period 2, the greater number of protons and valence electrons in the atom leads to a larger energy gap between the 2s and 2p orbitals and therefore less s–p mixing. As a result, the orbital cross-over observed in N_2 is also seen in B_2 and C_2 but not in O_2 and F_2. Similarly, in passing down a group there is a larger energy gap between valence s and p orbitals, leading to less s–p hybridisation.

In Figure 5.3 the radial function of the 1s orbital of two hydrogen atoms a distance of 1.4 Bohr radii apart are added together and squared to give the probability density of the bonding orbital, and subtracted and squared to give the probability density of the antibonding orbital. This is the distance between H atoms in an H_2 molecule [39]. The probability density of the molecular orbitals along the internuclear axis shows that most of the electron density is between the nuclei, consistent with Section 5.1. Note that in the antibonding orbital there is a nodal plane perpendicular to the internuclear axis midway between the hydrogen nuclei.

Figure 5.4 shows the probability density of the bonding and antibonding molecular orbitals in H_2 in the xy-plane using contour lines. It makes clear the nodal plane in the antibonding orbital.

Figures 5.5 and 5.6 show the molecular orbital probability densities for σ overlap between p orbitals in two carbon atoms a distance of 2.5 Bohr radii apart. This is the distance between two carbon atoms in an ethene molecule [39]. Again, there is a nodal plane perpendicular to the nuclear axis midway between the nuclei in the antibonding orbital.

The wavefunctions used in Figures 5.5 and 5.6 for carbon atoms are hydrogen-like but with Z corrected for the extra protons in the nucleus and the shielding by other electrons. Following Slater's rules in Section 3.3, Z_{eff} for the carbon 2p electron is $6 - 2.75 = 3.25$.

Figure 5.7 shows the molecular orbital probability densities for π overlap between p orbitals in two carbon atoms a distance of 2.5 Bohr radii apart. As in all molecular orbitals from π overlap, there is a nodal plane in

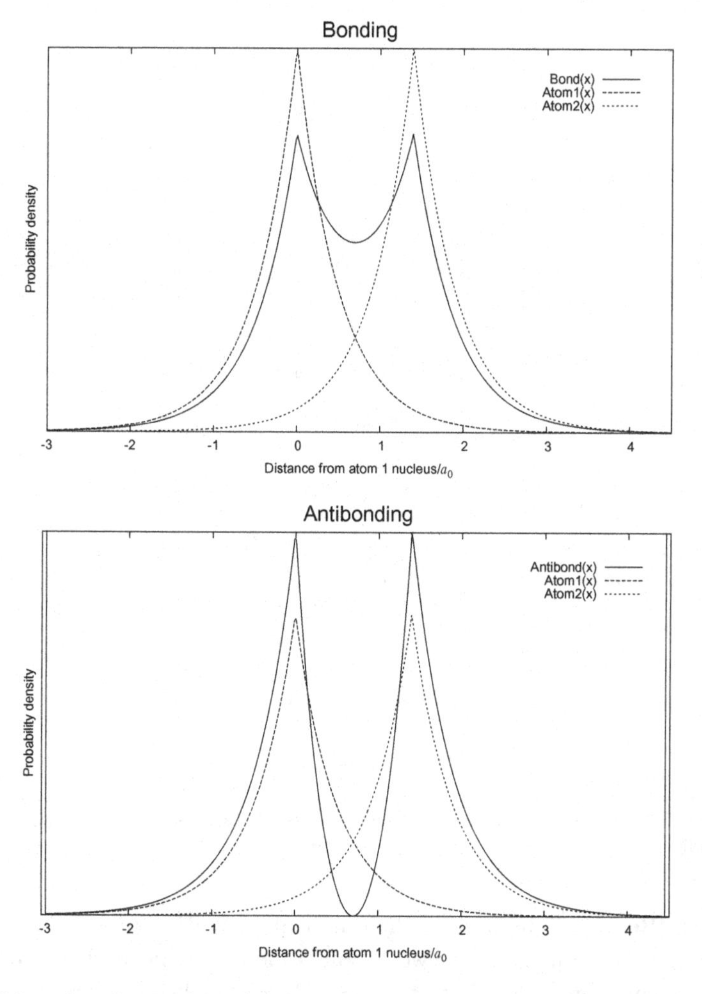

Figure 5.3. σ bonding and antibonding LCAOs of 1s orbitals in the H_2 molecule: probability density along the internuclear axis.

the plane of the molecule. In the antibonding orbital there is a further nodal plane perpendicular to the internuclear axis, as is seen in the σ antibonds. Note that the bonding molecular orbital is more compact than the antibonding MO.

More exotic forms of orbital overlap are also possible. When two transition metal atoms are in the appropriate geometry, it is possible for all four lobes of a d orbital to overlap with all four lobes of another d orbital. This is known as a δ bond. A δ bond was characterised in 1964 in the

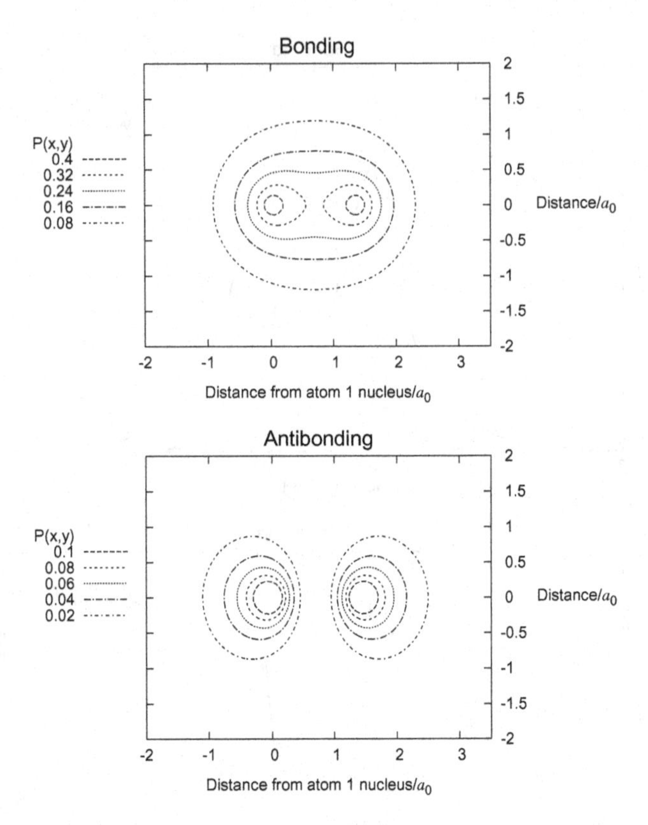

Figure 5.4. σ bonding and antibonding LCAOs of 1s orbitals in the H_2 molecule: probability density contour maps in the xy-plane.

complex $[Re_2Cl_8]^{2-}$ as part of the quadruple Re–Re bond (together with a σ and two π bonds) [77]. Two δ bonds have even been characterised as part of a quintuple Cr–Cr bond [78]. There has been a recent report of a ϕ interaction involving the overlap of all 6 lobes of a thorium 5f orbital that has lobes only in the xy-plane (see, for example, the $f_{x(x^2-3y^2)}$ and $f_{y(3x^2-y^2)}$ orbitals in Figure 2.5) together with those of ligands above and below the plane [79].

Indeed, f orbitals can exhibit σ, π, δ and ϕ bonding interactions, illustrated in Figure 5.8. All four types of overlap are seen in the thorium complex, though in that case the f orbitals overlap with different molecular orbitals in the cycloocta-1,3,5,7-tetraenyl ligands. Since it is a sandwich compound, the overlaps shown in Figure 5.8 show the f orbitals overlapping with neighbouring atoms' orbitals on either side. When there is

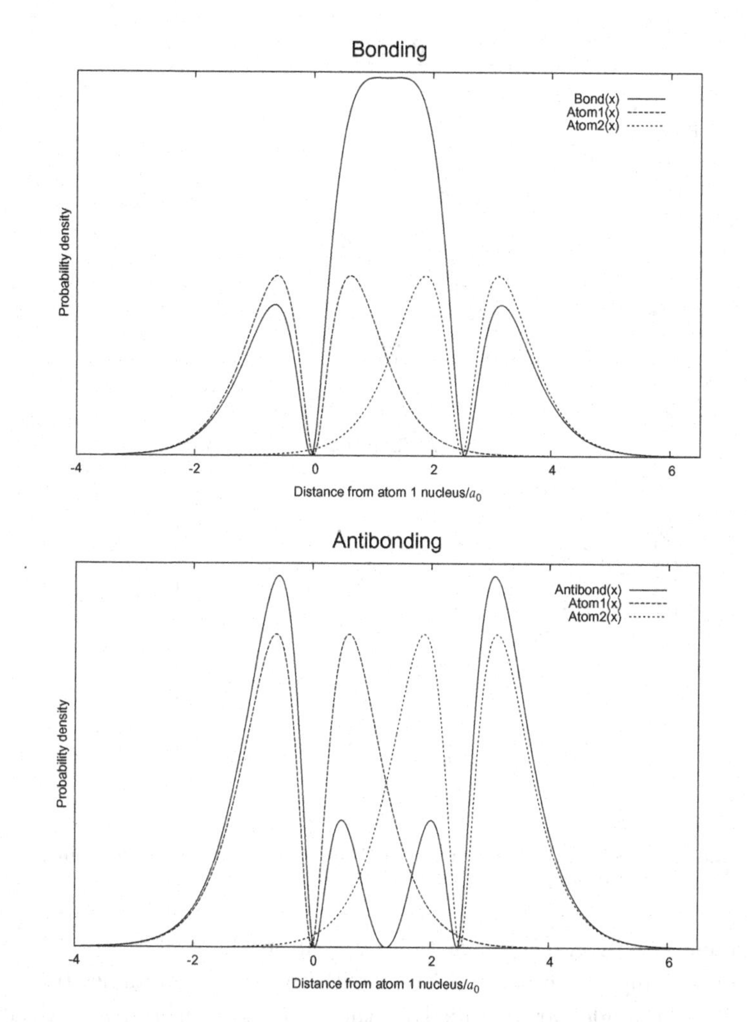

Figure 5.5. σ bonding and antibonding LCAOs of $2p_x$ orbitals: probability density along the internuclear axis.

π delocalisation (see Section 5.6), p_z orbitals overlap with their neighbours on either side, but usually, of course, orbitals just overlap with a single neighbour when there is a bonding interaction.

5.4 Hybridised Atomic Orbitals

The bond angles in differently hybridised molecules can be predicted using Valence Shell Electron Pair Repulsion theory, and have been determined

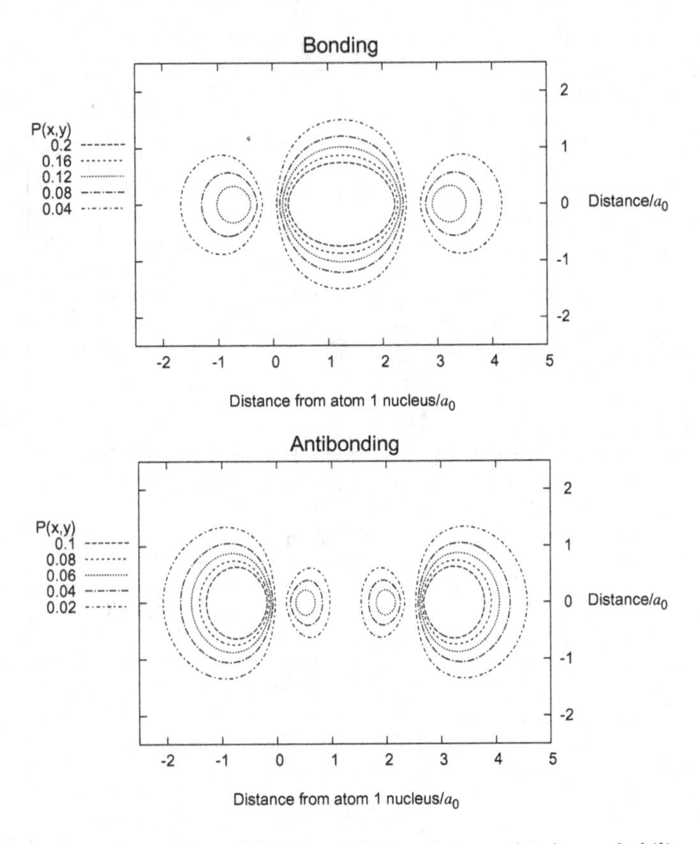

Figure 5.6. σ bonding and antibonding LCAOs of $2p_x$ orbitals: probability density contour maps in the xy-plane.

experimentally by X–ray diffraction (109° in methane, 120° in BF_3, etc.). These bond angles are not consistent with the 90° bond angles between p orbitals. The angles are justified by the concept of orbital hybridisation, popular with organic chemists in particular. s and p orbitals are considered to 'mix' to give these hybrids. It should be pointed out that such an approach, while a useful way of considering the electronic ground state of molecules, is an oversimplification. This is because hybridised orbitals (when relating to a platonic shape) are constructed to be energetically equivalent, yet photo-electron spectroscopy (PES) reveals different energy levels relating to the atomic orbitals that are distinguishable under the symmetry of the molecule. For example, the bonding electrons in methane have been shown to have two different energy levels (relating

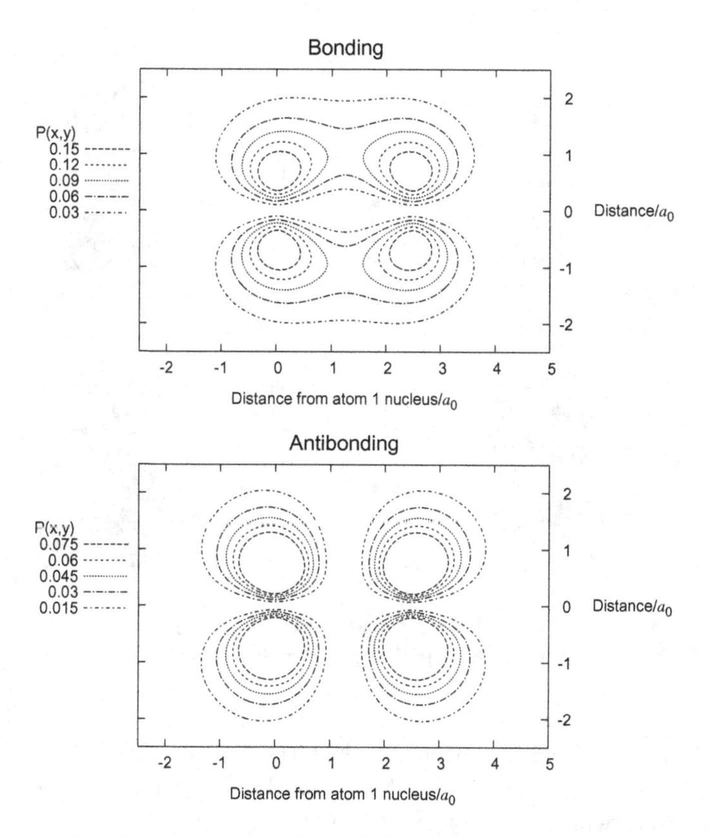

Figure 5.7. π bonding and antibonding LCAOs of $2p_y$ orbitals: probability density contour maps in the xy-plane.

to the separate bonding interactions of the s and p orbitals on the carbon), despite the apparent similarity of the four sp^3 hybrid orbitals on carbon [80].

Can the mixing of orbitals on an atom be justified on quantum mechanical grounds? We saw in Section 1.4 that we may be able to take linear combinations of the eigenfunctions with arbitrary constants in front of each eigenfunction, with the result still being an eigenfunction of the operator. It turns out that we can take useful linear combinations that are still eigenfunctions of the Hamiltonian operator.

Following the comparison of the addition of eigenfunctions to the addition of vectors in Section 1.4, we can consider the p orbitals as vectors along their given axis. Given that the eigenfunctions are all normalised,

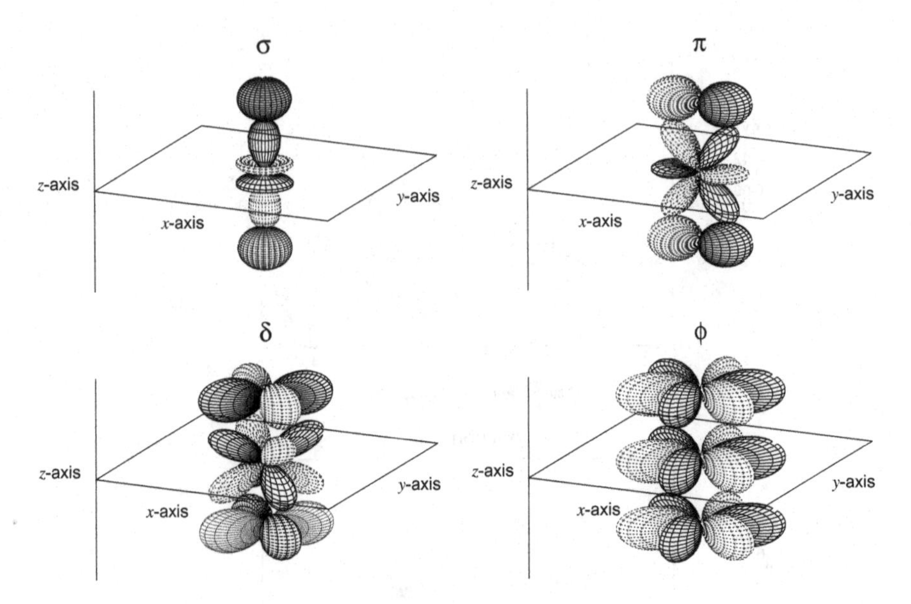

Figure 5.8. σ, π, δ and ϕ overlap geometries for f_{z^3}, f_{xz^2}, f_{xyz}, and $\mathrm{f}_{y(3x^2-y^2)}$ with s, p_x, d_{xy} and $\mathrm{f}_{y(3x^2-y^2)}$ orbitals, respectively, above and below.

we can assume that the p orbital vectors are all the same length. The hybridisation of a carbon atom involves the linear combination of the s orbital with $(3 - n)$ p orbitals where n is the number of π bonds on the carbon atom (which may not exceed two).

In the case of sp^3 hybridisation, four bonds around the carbon atom point to the corners of a tetrahedron. If we consider the carbon nucleus to be at the centre of a cube, then the bonds would point to four corners of the cube that are mutually diagonally opposite each other on each face. This geometry is shown in Figure 5.9. The bonds are shown as vectors representing the contribution of each p orbital vector, **x**, **y** and **z**, to the sp^3 hybrid.

Considering the p orbitals to be vectors whose length is equal to half the length of an edge of the cube, we can construct the four hybrid orbitals as:

$$\mathrm{sp}^3(1) = N_s\mathrm{s} + N_p\mathrm{p}_x + N_p\mathrm{p}_y + N_p\mathrm{p}_z \tag{5.17a}$$

$$\mathrm{sp}^3(2) = N_s\mathrm{s} + N_p\mathrm{p}_x - N_p\mathrm{p}_y - N_p\mathrm{p}_z \tag{5.17b}$$

$$\mathrm{sp}^3(3) = N_s\mathrm{s} - N_p\mathrm{p}_x + N_p\mathrm{p}_y - N_p\mathrm{p}_z \tag{5.17c}$$

$$\mathrm{sp}^3(4) = N_s\mathrm{s} - N_p\mathrm{p}_x - N_p\mathrm{p}_y + N_p\mathrm{p}_z, \tag{5.17d}$$

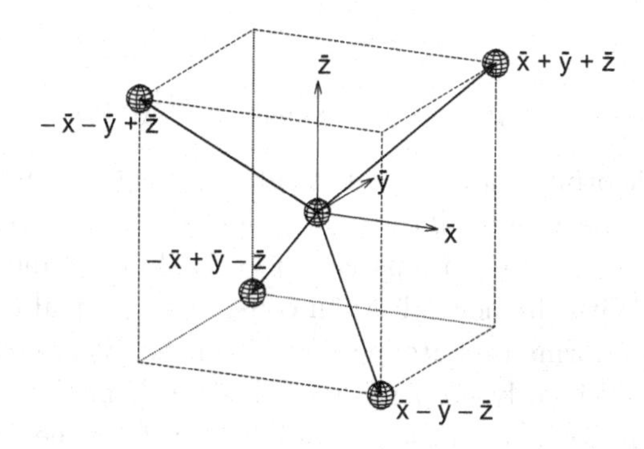

Figure 5.9. The sp^3 bonds represented as vectors in terms of p orbital vectors.

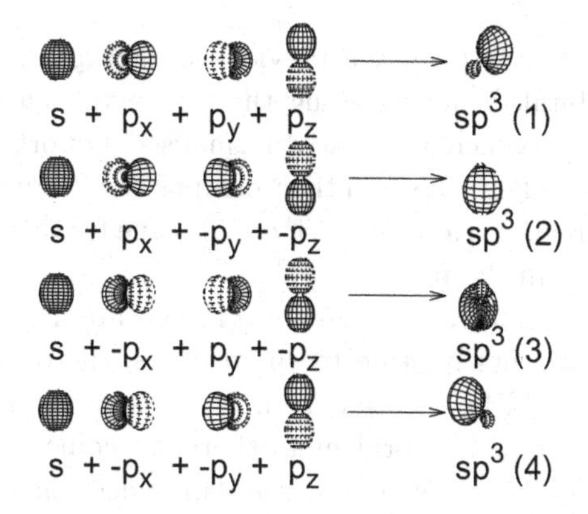

Figure 5.10. sp^3 hybrid orbitals as the linear combination of s and p orbitals shown in equations (5.17).

where the N represent normalisation constants. Since the x-, y- and z-axes are equivalent in a tetrahedron, the normalisation constants in front of each p orbital are assumed to be equal. Figure 5.10 shows how the shape of the sp^3 hybrid orbitals follows from the linear combinations given in equations (5.17).

As well as being normalised, the hybrid orbitals should also be orthogonal to each other. To be orthogonal, the scalar (dot) product of any two of the sp^3 hybrids should come to zero. Any cross terms in the multiplication come to zero since the atomic orbitals are all defined to be orthogonal with one another. The dot product of any two different sp^3 hybrids gives the same expression from which the normalisation constants may be found:

$$\langle \mathrm{sp}^3 | \mathrm{sp}^{3\,\prime} \rangle = 0 = N_s^2 \langle \mathrm{s}|\mathrm{s} \rangle - N_p^2 \langle \mathrm{p}|\mathrm{p} \rangle. \tag{5.18}$$

Since the atomic orbitals are normalised all the integrals come to 1, leaving $N_s^2 = N_p^2$. Normalisation of all the hybrids leads to the common expression, $N_s^2 + 3N_p^2 = 1$. These equations can be solved simultaneously, yielding $N_s = N_p = \frac{1}{2}$. With the normalisation constants all equal to $\frac{1}{\sqrt{n}}$, where n is the number of normalised atomic orbitals in the linear combination, the hybrid orbital is evidently properly normalised. It is also evident that the s orbital makes up a quarter of the probability density of the orbital, which is its fraction of the 4-orbital hybrid. The three p orbitals are equally weighted in the wavefunction as they are symmetrically equivalent in a tetrahedral environment.

Following the results from the previous paragraph, it is obvious that two sp hybrid orbitals pointing along the z-axis will have wavefunctions $|\mathrm{sp}\rangle = \frac{1}{\sqrt{2}}(|\mathrm{s}\rangle \pm |\mathrm{p}_z\rangle)$, which are clearly normalised and orthogonal with one another. These hybrid orbitals and their composition from s and p_z orbitals are shown in Figure 5.11. Note that the nodal plane for these hybrid orbitals does not pass through the nucleus.

The wavefunctions for the sp^2 hybrid orbitals are not quite so obvious. Following the arguments relating to sp^3 orbitals, the coefficient for the s orbital component will be $\frac{1}{\sqrt{3}}$. The sum of the squares of the coefficients for the p orbitals in each hybrid must therefore come to 2/3. We define the three sp^2 orbitals to be in the xy-plane such that one is pointing

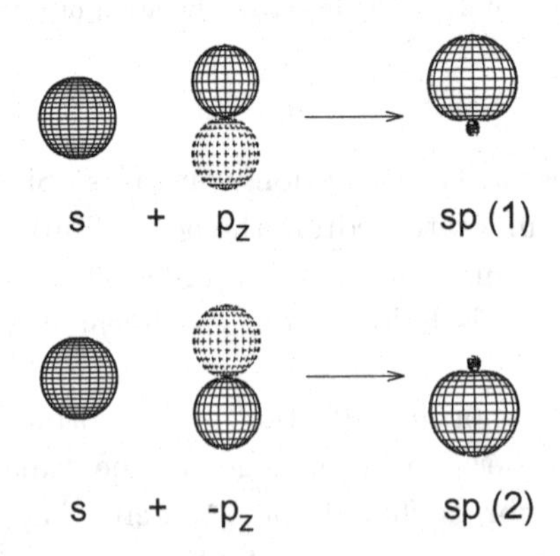

Figure 5.11. sp hybrid orbitals as the linear combination of s and p_z orbitals.

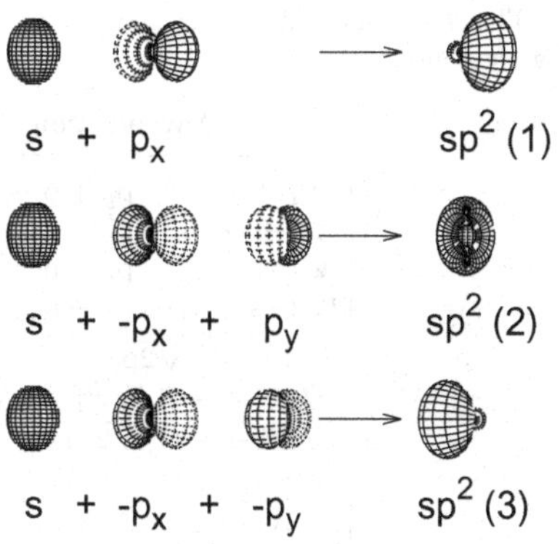

Figure 5.12. sp^2 hybrid orbitals as the linear combination of s and p orbitals shown in Table 5.1.

along the positive x-axis, as shown in Figure 5.12. The hybrid orbital pointing along the x-axis will only have the p_x orbital combined with the s orbital. We can therefore immediately deduce that the coefficient in front of the p_x orbital in that hybrid is $\sqrt{2/3}$. In the other two sp^2 hybrid orbitals the coefficient in front of the p_x orbital must be negative and, due to the angles between the hybrids, will have the factor $\cos 60° = \frac{1}{2}$. The normalisation constant in front of the p_y orbital will be positive in one of these hybrids and negative in the other. The angles dictate that a factor of $\cos 30° = \sqrt{3}/2$ appears with the p_y orbitals. Squaring and summing these trigonometric factors gives 1. As they account for two p orbitals the squares of these factors should sum to 2. This is achieved by multiplying these trigonometric factors by $\sqrt{2}$. Multiplying the trigonometric factors by the overall normalisation factor of $\frac{1}{\sqrt{3}}$ will therefore give the correct overall normalisation constants, which are collected in Table 5.1.

Exercise 68

Show that the $sp^2(1)$ and $sp^2(2)$ orbitals are orthogonal by calculating $\langle sp^2(1)|sp^2(2)\rangle$. Use the wavefunctions in Table 5.1.

The geometry of the linear combinations for the sp^2 hybrid orbitals is shown in Figure 5.12, with the lobes drawn with unbroken lines having a positive phase. It is assumed that the s orbital has a positive phase, but this is not the case for 2s orbitals, where its radial node means that the

Table 5.1. The wavefunctions of all the sp^n orbitals for second-row elements.

Hybrid orbital	Wavefunction
$sp^3(1)$	$(1/2)(-s + p_x + p_y + p_z)$
$sp^3(2)$	$(1/2)(-s + p_x - p_y - p_z)$
$sp^3(3)$	$(1/2)(-s - p_x + p_y - p_z)$
$sp^3(4)$	$(1/2)(-s - p_x - p_y + p_z)$
$sp^2(1)$	$(1/\sqrt{3})(-s + \sqrt{2}p_x)$
$sp^2(2)$	$(1/\sqrt{3})(-s - (1/\sqrt{2})p_x + \sqrt{3/2}p_y)$
$sp^2(3)$	$(1/\sqrt{3})(-s - (1/\sqrt{2})p_x - \sqrt{3/2}p_y)$
$sp(1)$	$(1/\sqrt{2})(-s + p_z)$
$sp(2)$	$(1/\sqrt{2})(-s - p_z)$

exterior of the orbital (which will be overlapping with other orbitals) has a negative phase. Given that second-row elements most commonly exemplify these hybrid orbitals, the normalisation constant for the s orbital will have to be made negative so that the combinations of atomic orbitals follow the scheme in Figure 5.12. Note that the main lobes appear quite 'fat'; the 'back-tail' lobes are small but not negigible as far as rearward interactions are concerned.

Figure 5.13 shows the amplitude functions in the xy-plane using contour lines for the three sp^2 orbitals in a carbon atom using the linear combinations of atomic orbitals given in Table 5.1.

5.5 Hybridised Bonding and Antibonding Orbitals

BF_3 is a classic example of a molecule with an sp^2 hybridised central atom. Using the rule in Section 5.4 one would expect the fluorine atoms to be sp^3 hybridised. However they may be considered to be sp^2 hybridised since there is also some π bonding between a lone pair in a p orbital on each fluorine and the vacant p orbital on boron [81]. The average B–F bond energy in BF_3 is greater than any known single bond, at 646 kJ mol^{-1}; the B–F bond length is also surprisingly short [81], shorter than the C–F bond despite the smaller atomic radius of carbon.

Figure 5.14 shows the bonding and antibonding molecular orbitals in BF_3 from overlap of the sp^2 hybrid orbitals on boron and fluorine. This gives a reasonable approximation of the ground state of the molecule, though strictly, in the LCAO approach, the boron s and p orbitals should

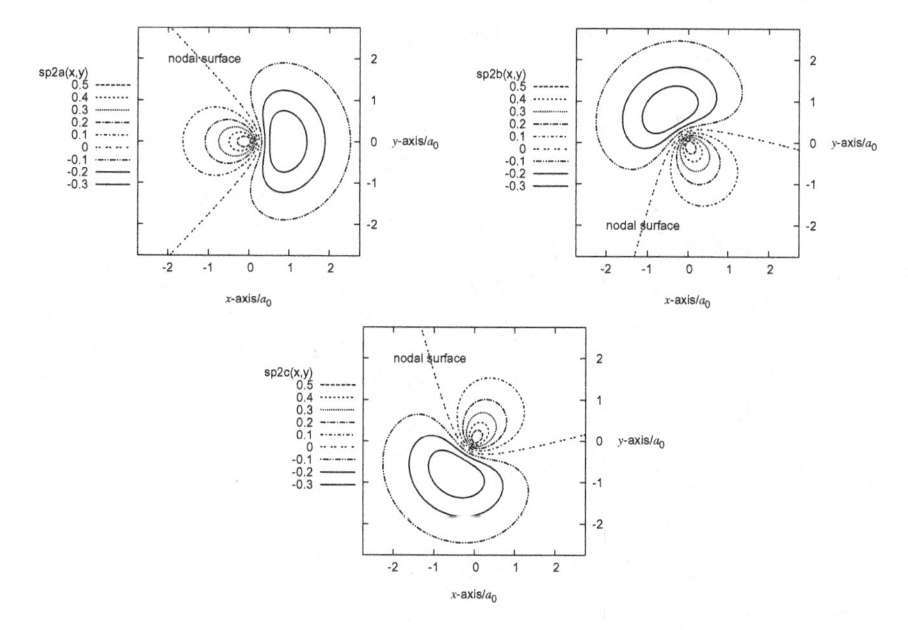

Figure 5.13. The three sp^2 hybrid orbitals: amplitude function contour maps in the xy-plane.

combine directly with suitable combinations of the fluorine atoms' atomic orbitals. The plot uses contour lines to visualise the probability density in the plane of the molecule. Electron density from the inner electrons on B and F, and from the π interactions between boron and fluorine are not shown.

In the full molecular orbital treatment of a polyatomic molecule, all the atomic orbitals on all the atoms in the molecule need to be taken together, with their group theoretical label to determine which orbitals can interact. In sophisticated treatments, strongly deformed atomic orbitals are used in the linear combinations, with different radial and angular functions from those used in hydrogen atomic orbitals. The overlap of hybridised orbitals is therefore a simplification, but very often a useful one to gain a reasonable general impression of the bonding.

Exercise 69

Show that the sp(1) and sp(2) orbitals are eigenfunctions of the Hamiltonian (energy) operator shown in equation (2.22). Use the wavefunctions in Table 5.1.

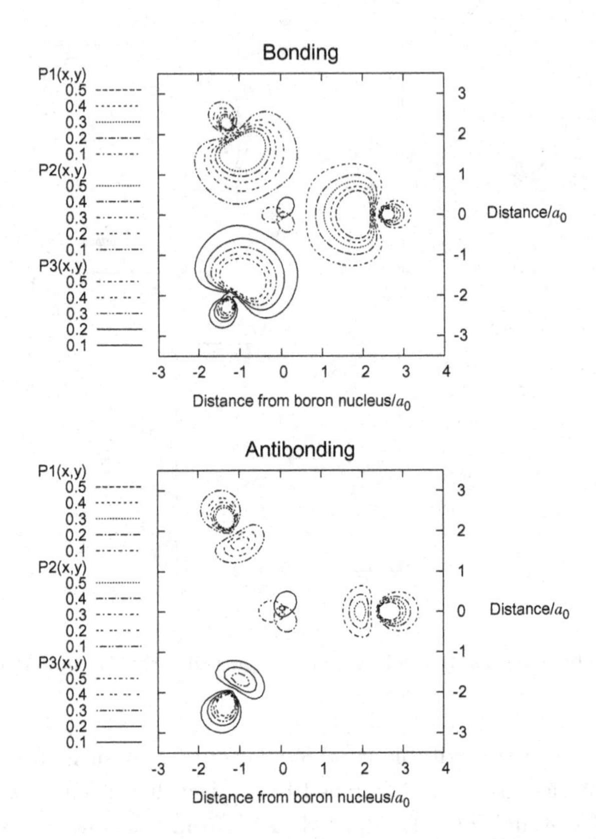

Figure 5.14. The three σ bonds and the three σ antibonds of BF_3 formed from overlap of sp^2 hybrid orbitals on boron and fluorine: probability density contour maps in the xy-plane.

5.6 π Bonding in Benzene

With benzene being flat, the singly occupied p_z orbital on each carbon can overlap in π geometry with its neighbours on either side. It is commonly envisaged that the six π bonding electrons (one from each carbon) can be represented as in Figure 5.15. However, that figure represents a single molecular orbital — the π ground-state MO — and, as such, can only accommodate two electrons.

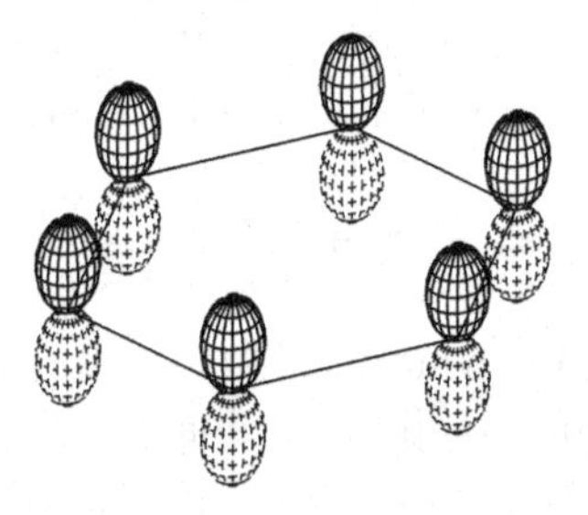

Figure 5.15. The relative phases of the atomic p orbital amplitude functions in the lowest-energy π molecular orbital of benzene.

Exercise 70

On symmetry grounds, we can assume that the coefficients in front of the atomic p_z wavefunctions are the same in the ground-state π MO of benzene. Write out the normalised wavefunction of this MO.

Given that we include six p_z orbitals in the MO wavefunction, they must combine to give a total of six MOs. In order to find their wavefunctions and energies we need to solve the secular determinant for this LCAO following the methods of Section 5.2. To simplify the algebra we neglect the overlap integrals between different p_z orbitals, S_{ij} where $i \neq j$. To simplify the notation we divide through the secular equations by β and equate $\frac{\alpha - E}{\beta}$ to x. This leaves us with a simple-looking determinant to solve:

$$\begin{vmatrix} x & 1 & 0 & 0 & 0 & 1 \\ 1 & x & 1 & 0 & 0 & 0 \\ 0 & 1 & x & 1 & 0 & 0 \\ 0 & 0 & 1 & x & 1 & 0 \\ 0 & 0 & 0 & 1 & x & 1 \\ 1 & 0 & 0 & 0 & 1 & x \end{vmatrix} = 0. \tag{5.19}$$

To expand $n \times n$ determinants that are larger than 2×2 they are reduced one unit at a time by taking each element along a row (or column) and multiplying it by its minor, which is defined to be the $(n-1) \times (n-1)$ determinant omitting the row and column of the element it's multiplied by. There is also a phase factor, $(-1)^{i+j}$, where i is the row number and j is the column number.

$$\begin{vmatrix} a & b & c \\ d & e & f \\ g & h & i \end{vmatrix} = a\begin{vmatrix} e & f \\ h & i \end{vmatrix} - b\begin{vmatrix} d & f \\ g & i \end{vmatrix} + c\begin{vmatrix} d & e \\ g & h \end{vmatrix} \tag{5.20}$$

Expanding the determinant in equation (5.19) leads to an equation in x that goes up to the sixth power, and so isn't easy to solve:

$$x^6 - 6x^4 + 9x^2 - 4 = 0. \tag{5.21}$$

Another method of solving equation (5.19) is to simplify the determinant using two rules that apply to all determinants.

(1) A column (or row) can be added to another column (or row) without changing the determinant.
(2) If there is a common factor in all the elements in a column (or row) then this factor may be taken out as a factor by which the whole determinant may be multiplied.

Using these rules, the six solutions to equation (5.19) are found to be $x = \pm 1(\text{twice}), \pm 2$.

Exercise 71

Apply the rules for simplifying determinants to solve equation (5.19).

These solutions for x translate into the energies

$$E = \alpha \pm \beta(\text{twice}), \quad \alpha \pm 2\beta,$$

shown in the MO scheme in Figure 5.16. It is instructive to compare this to the result from the simplest organic π system, in ethene. Using the approach taken for benzene, ethene gives the following determinantal equation:

$$\begin{vmatrix} x & 1 \\ 1 & x \end{vmatrix} = (x+1)(x-1) = 0. \tag{5.22}$$

This equation is evidently solved by $x = \pm 1$, i.e. $E = \alpha \pm \beta$. Four of the six solutions for the benzene MOs are equivalent to these, but the ground state is another unit of β lower in energy, reflecting the stabilisation from the delocalisation of the π electrons over all six carbons, and manifest in the additional thermodynamic stability of benzene compared with simple alkenes.

In order to determine the coefficients in the MOs, the energy for each MO can be inserted back in to the secular equations. We have already

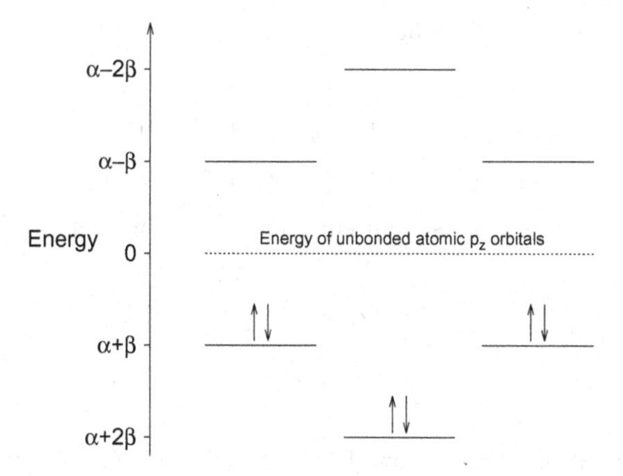

Figure 5.16. The molecular orbital scheme for benzene.

proposed a wavefunction for the ground state,

$$\psi_{\mathrm{MO,GS}} = \frac{1}{\sqrt{6}}(\psi_{p_z}(1) + \psi_{p_z}(2) + \psi_{p_z}(3) + \psi_{p_z}(4) + \psi_{p_z}(5) + \psi_{p_z}(6)).$$

If we insert $E = \alpha + 2\beta$ into the six secular equations we find the following relations between the coefficients:

$$\begin{aligned} -2c_1 + c_2 + c_6 &= 0 \quad c_3 - 2c_4 + c_5 = 0 \\ c_1 - 2c_2 + c_3 &= 0 \quad c_4 - 2c_5 + c_6 = 0 \\ c_2 - 2c_3 + c_4 &= 0 \quad c_5 - 2c_6 + c_1 = 0. \end{aligned} \tag{5.23}$$

We can say by inspection that all values of c_i being equal is an acceptable solution, which supports our proposed wavefunction. We find closely related equations for the highest energy MO, with $E = \alpha - 2\beta$:

$$\begin{aligned} 2c_1 + c_2 + c_6 &= 0 \quad c_3 + 2c_4 + c_5 = 0 \\ c_1 + 2c_2 + c_3 &= 0 \quad c_4 + 2c_5 + c_6 = 0 \\ c_2 + 2c_3 + c_4 &= 0 \quad c_5 + 2c_6 + c_1 = 0. \end{aligned} \tag{5.24}$$

We can say by inspection that an acceptable solution is for all values of c_i to have the same magnitude but to alternate in sign. This implies that every orbital is out of phase with its neighbour, which is consistent with the MO being the most antibonding one. This corresponds to the wavefunction

$$\psi_{MO} = \frac{1}{\sqrt{6}}(\psi_{p_z}(1) - \psi_{p_z}(2) + \psi_{p_z}(3) - \psi_{p_z}(4) + \psi_{p_z}(5) - \psi_{p_z}(6)).$$

The situation is more complicated for the other four MOs since the two remaining energies are doubly degenerate. Inserting $E = \alpha + \beta$ into the secular equations we find the following relations between the coefficients:

$$\begin{aligned} -c_1 + c_2 + c_6 &= 0 \quad c_3 - c_4 + c_5 = 0 \\ c_1 - c_2 + c_3 &= 0 \quad c_4 - c_5 + c_6 = 0 \\ c_2 - c_3 + c_4 &= 0 \quad c_5 - c_6 + c_1 = 0. \end{aligned} \tag{5.25}$$

If we add adjacent expressions from equations (5.25) we get the following results:

$$\begin{aligned} c_3 + c_6 &= 0 \\ c_2 + c_5 &= 0 \\ c_4 + c_1 &= 0. \end{aligned} \tag{5.26}$$

The results of equations (5.26) show that the coefficients of opposite carbon atoms in the ring are equal but of opposite sign. This is consistent with a single nodal plane in the molecule, i.e. a plane bisecting molecule with AOs of opposite phase on either side. The fact that there are two degenerate MOs of this energy implies that there are two different positions of the nodal plane.

Inserting $E = \alpha - \beta$ into the secular equations we find the following relations between the coefficients:

$$\begin{aligned} c_1 + c_2 + c_6 &= 0 \quad c_3 + c_4 + c_5 = 0 \\ c_1 + c_2 + c_3 &= 0 \quad c_4 + c_5 + c_6 = 0 \\ c_2 + c_3 + c_4 &= 0 \quad c_5 + c_6 + c_1 = 0. \end{aligned} \tag{5.27}$$

If we add adjacent expressions from equations (5.25) we get the following results:

$$\begin{aligned} c_3 - c_6 &= 0 \\ c_2 - c_5 &= 0 \\ c_4 - c_1 &= 0. \end{aligned} \tag{5.28}$$

The results of equations (5.28) show that the coefficients of opposite carbon atoms in the ring are equal. However, the secular equations (5.27) do not permit all the coefficients to be the same sign, so opposite carbons could be the same sign on account of two nodal planes bisecting the ring between them.

The highest energy MO with all carbon AOs out of phase with their neighbours is the result of three nodal planes. Hence the four energy levels of the benzene π system seem to be characterised by their number of nodal planes: 0, 1, 2 and 3.

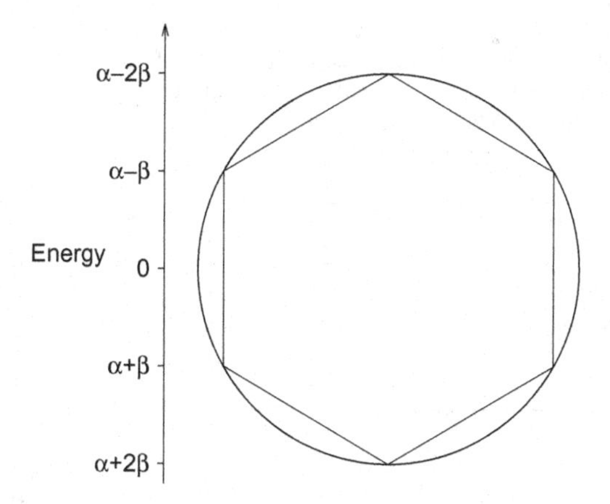

Figure 5.17. The Frost diagram for benzene.

The energies of the π MOs in benzene are not only simple expressions but also show a geometric pattern that reflects the molecule's shape. This is made clear in the Frost diagram shown in Figure 5.17. A regular hexagon, as is produced by the carbon atoms in benzene, is placed inside a circle with a vertex placed at the bottom of the circle. The energy of the six MOs can be read off from the position of each vertex, given that α is at half the height of the circle and $\alpha + 2\beta$ is at the bottom.

Indeed the Frost diagram is equally predictive for all cyclic π systems. All cyclic π systems have a singly degenerate ground state with no nodal planes. Higher energy levels are doubly degenerate. The highest-energy state is therefore only singly degenerate when there is an even number of atoms in the ring; otherwise it is doubly degenerate.

The simple solutions and geometric relationships involved with the benzene π system suggest there must be a simpler method to determine the energies and coefficients in MOs than the protracted simplification of determinants or solving equations with high powers. Coulson has published methods involving the properties of determinants and much algebra to give the general energy expressions for chemically-relevant determinants [82]. Methods that are perhaps more elegant have been published by Burdett [83].

The cyclic Frost diagram suggests that the MO scheme is essentially a two-dimensional problem and so complex numbers should be able to offer a succinct solution. The complex exponential $e^{i\theta}$ plots a circle in the Argand diagram as θ goes from 0 to 2π. The choice of a complex exponential is

Table 5.2. The value of the coefficients c_{pj} in the absence of the normalisation constant for each of the p atomic orbitals in each j molecular orbital in benzene following equation (5.29). $\epsilon = \exp(2\pi i/6)$.

j	p_1	p_2	p_3	p_4	p_5	p_6
0	1	1	1	1	1	1
+1	1	ϵ	$-\epsilon^*$	-1	$-\epsilon$	ϵ^*
−1	1	ϵ^*	$-\epsilon$	-1	$-\epsilon^*$	ϵ
+2	1	$-\epsilon^*$	$-\epsilon$	1	$-\epsilon^*$	$-\epsilon$
−2	1	$-\epsilon$	$-\epsilon^*$	1	$-\epsilon$	$-\epsilon^*$
3	1	-1	1	-1	1	-1

justified by group theory where it is the generator of cyclic groups. The following general expression is for the jth MO wavefunction of an n-atom cyclic π system [83]. j takes the sequence 0, ±1, ±2, 3 in benzene. It is the number of nodal planes, taking the negative value in addition to the positive one for doubly degenerate levels:

$$\psi_{\mathrm{MO},j} = \sum_{p=1}^{n} c_{pj}\psi_{\mathrm{AO},p} = \frac{1}{\sqrt{n}}\sum_{p=1}^{n}\left[\exp\left(\frac{2\pi i j(p-1)}{n}\right)\right]\psi_{\mathrm{AO},p}. \tag{5.29}$$

In equation (5.29) the variable p counts the n atomic p orbitals that overlap to form the molecular orbitals. The value of the coefficients c_{pj} in the absence of the normalisation constant for each of the p atomic orbitals in each j molecular orbital is given in Table 5.2, where $\epsilon = \exp(2\pi i/6)$.

Table 5.2 shows that the coefficients c_{pj} are real for the singly degenerate states and complex for the doubly degenerate ones. Purely real or purely imaginary linear combinations can be made by taking the sum or difference of the $+j$ and $-j$ wavefunctions, respectively. (Purely imaginary is satisfactory as when multiplied by its complex conjugate it will be purely real.) Using these linear combinations, all the benzene π MOs are given in Table 5.3.

The 12 in the normalisation constants arises because $\psi(1)$ and $\psi(4)$, when squared in the normalisation process, each contribute 4 towards the sum of overlap integrals before that sum is square rooted. The form of the MOs is consistent with the position of the nodal planes shown in the molecular orbitals shown in Figure 5.18. Taking the difference between the two complex j wavefunctions leads to the disappearance of two of the AOs in the wavefunction. This gives the second position of the nodal plane, running over carbons 1 and 4, as seen in Figure 5.18.

Table 5.3. The benzene π MO wavefunctions with normalisation constant N.

j label	N	Linear combination of AOs
$\|0\rangle$	$1/\sqrt{6}$	$\psi(1)+\psi(2)+\psi(3)+\psi(4)+\psi(5)+\psi(6)$
$\|1\rangle+\|-1\rangle$	$1/\sqrt{12}$	$2\psi(1)+\psi(2)-\psi(3)-2\psi(4)-\psi(5)+\psi(6)$
$\|1\rangle-\|-1\rangle$	$i/2$	$\psi(2)+\psi(3)-\psi(5)-\psi(6)$
$\|2\rangle+\|-2\rangle$	$1/\sqrt{12}$	$2\psi(1)-\psi(2)-\psi(3)+2\psi(4)-\psi(5)-\psi(6)$
$\|2\rangle-\|-2\rangle$	$i/2$	$\psi(2)-\psi(3)+\psi(5)-\psi(6)$
$\|3\rangle$	$1/\sqrt{6}$	$\psi(1)-\psi(2)+\psi(3)-\psi(4)+\psi(5)-\psi(6)$

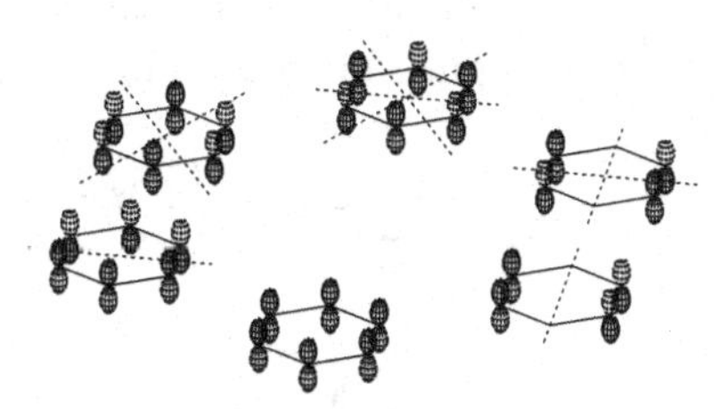

Figure 5.18. The combinations of atomic p_z orbitals that make up the π molecular orbitals on benzene.

An increase in the number of nodal planes causes an increase in the curvature of the wavefunction and therefore, due to the operator for kinetic energy involving a second derivative, an increase in the kinetic energy associated with the orbital.

A general expression for the energy of the bonding interaction in cyclic π systems can be derived easily from the generalised wavefunction of equation (5.29). This is achieved by summing the interactions of an atomic p orbital with label p with its neighbours on either side, $p-1$ and $p+1$, and summing over p [83]. The interaction is expressed in terms of the coefficients c_{pj} since the atomic wavefunctions are taken to be normalised:

$$E_j = \alpha + \beta \sum_{p=1}^{n} (c_{pj}^* c_{(p+1)j} + c_{pj}^* c_{(p-1)j}). \tag{5.30}$$

We then substitute into equation (5.30) the exponentials from equation (5.29) that express c_{pj} in terms of $(p-1)$, remembering to change the sign of the imaginary exponent in the complex conjugates:

$$E_j = \alpha + \beta \sum_{p=1}^{n} \left\{ \frac{1}{\sqrt{n}} \exp\left(\frac{-2\pi \mathrm{i} j(p-1)}{n}\right) \frac{1}{\sqrt{n}} \exp\left(\frac{2\pi \mathrm{i} jp}{n}\right) \right.$$
$$\left. + \frac{1}{\sqrt{n}} \exp\left(\frac{-2\pi \mathrm{i} j(p-1)}{n}\right) \frac{1}{\sqrt{n}} \exp\left(\frac{2\pi \mathrm{i} j(p-2)}{n}\right) \right\}$$
$$= \alpha + \beta \sum_{p=1}^{n} \frac{1}{n} \left\{ \exp\left(\frac{2\pi \mathrm{i} j}{n}\right) + \exp\left(\frac{-2\pi \mathrm{i} j}{n}\right) \right\}. \tag{5.31}$$

Equation (5.31) has given us a sum of two exponentials where the imaginary exponents differ only in their sign. This results in a real cosine function,

$$E_j = \alpha + \beta \sum_{p=1}^{n} \frac{1}{n}.2\cos\left(\frac{2\pi j}{n}\right)$$
$$= \alpha + 2\beta \cos\left(\frac{2\pi j}{n}\right), \tag{5.32}$$

where $j = 0, \pm 1, \pm 2 \ldots$.

In the case of benzene, where $n = 6$, $j = 0$ gives the ground state energy, $E = \alpha + 2\beta$, and $j = 3$ gives the highest energy MO, $E = \alpha - 2\beta$. $j = \pm 1$ and $j = \pm 2$ give the two pairs of degenerate states. The Hückel $(4n+2)$ rule predicts the number of atoms in a cyclic π system that gives the greatest energetic stability, i.e. aromaticity. These are the rings for which all the π electrons in their ground state are paired in MOs with energies below their atomic energies. The Frost diagram makes it clear how the Hückel numbers confer the greatest energetic advantage to the π system.

The link between the value of j and the number of nodal planes in the cyclic π system MOs is relevant to the f orbital interactions discussed in Section 5.3. As can be seen in Figure 5.8 the different types of bond can be characterised by the number of nodes cutting across the plane perpendicular to the bond axis (the z-axis in Figure 5.8): no nodes for σ, 1 for π, 2 for δ and 3 for ϕ — following the quantum numbers that relate to these letter in Latin form. The reason that the f orbitals could bond in all these geometries with the $C_8H_8^{2-}$ cycloocta-1,3,5,7-tetraenyl ion is that, having eight carbon atoms in a ring, the π molecular orbitals could have zero, one, two, three or four nodal planes.

Modern extensions to these aromatic ideas include Möbius aromaticity [84] and concentric planar doubly π-aromatic clusters [85].

A major limitation to the bonding theory as presented so far is that there are no explicit electron–electron interactions, which are important for angular-momentum-carrying unpaired electrons. Each molecular orbital confers quantum numbers and symmetry labels to electrons in that orbital, and the individual electrons will couple to a resultant. For example, a simplistic interpretation of the benzene MO diagram of Figure 5.16 would predict that the lowest energy excitation involved an energy of -2β with higher energy excitations of -3β and -4β. The actual spectrum is more complex on account of electron–electron interactions. This additional interaction needs to be taken into account when interpreting the spectra of molecules and is considered in detail in [86]. Furthermore there will be selection rules to consider; these will be explored in the next chapter. The remainder of this book, however, will focus on atomic spectra.

Chapter 6

Atomic Spectroscopy

6.1 Electric-Dipole Transitions and Selection Rules

Light and other forms of electromagnetic radiation involve oscillating electric and magnetic fields. In principle either of these can excite electrons in dipole, quadrupole or higher-multipole transitions. The most important mechanism for exciting electrons is electric-dipole transitions, which will be the focus of this chapter.

The electric-dipole mechanism treats the electric-field component of the electromagnetic radiation as an oscillating vector, which transfers its energy and angular momentum to the electrons. This transfer occurs when there is resonance between the photon frequency, ν, and the energy gap, ΔE, between the initial and final states through $\Delta E = h\nu$. It is a requirement for resonance that the excitation of the electron involves a dipole change in the electron density. An excitation from a spherical s orbital to another spherical orbital involves no dipole change, but the excitation from an s orbital to a p orbital, for example, does. Curiously, no one-electron orbital has an electric dipole; however the superposition of two orbitals, involved in an electronic transition, can create one.

The critical quantity in determining the viability of an electric-dipole transition is the transition moment:

$$\mu_{if} = \langle \psi_i \mid \hat{\mu} \mid \psi_f \rangle.$$

The transition moment connects the initial wavefunction, ψ_i, and the final wavefunction, ψ_f, with the electric-dipole operator, $\hat{\mu} = e\hat{r}$, where e is the electron charge and $\hat{r}$ is the operator for the vector of the electric field. The transition moment is generally complex since it includes phase factors associated with the two states that give information about the polarisation of the transition. Its square modulus, $|\mu_{if}|^2$, gives the strength of the interaction and is proportional to the transition rate, following Fermi's golden rule [10].

If the transition moment is zero it is usually because the transition is forbidden by selection rules that relate to the symmetry of the states and the electric-dipole operator. In order for an electric-dipole transition to be observed, the dipole moment must be a scalar (a number rather than a vector, etc.) since the transition rate must also be a scalar.

Rule 1: Parity

The simplest symmetry element of relevance to selection rules is parity symmetry. Parity is symmetry with respect to a centre of inversion (which, for atoms, is the nucleus). If an orbital is unchanged by reflection through the nucleus then it is said to have even parity; if the phases of its lobes are reversed by such an operation, the orbital is said to have odd parity. Inspection of the atomic orbital plots in Figures 2.2, 2.3, 2.4 and 2.5 reveals alternating parity — even, odd, even, odd — which is equivalent to the positive and negative results of $(-1)^l$, where l is the orbital angular momentum quantum number of the orbital.

In order for an electric-dipole transition to be allowed by the parity selection rule, the product of the symmetries of both states and the operator must be even, since the transition moment, being a scalar, must have even symmetry. Thinking of the vector $\mathbf{r}$ as an arrow pointing from the origin to point r, it is clearly of odd parity symmetry. Hence, for the transition moment to have even parity, the initial and final states must have opposite parity. This is also known as the Laporte selection rule. It relates to the individual electron parities, not to many-electron coupled wavefunctions (as the latter are antisymmetrised).

$$\Delta l = \text{odd} \tag{6.1}$$

Exercise 72

Which electric-dipole transitions between subshells are allowed and forbidden by the parity selection rule? Comment on the implications for the colour of transition metal complex ions.

Curiously, the magnetic-dipole operator has even parity since a magnetic dipole is not a vector, but rather a pseudo vector as it is produced by a current loop. This means that the parity selection rules for magnetic-dipole transitions are the opposite of those for electric-dipole transitions.

Rule 2: Angular momentum

In order for the electric-dipole transition moment to be a scalar, i.e. fully symmetric, the angular momenta involved in the transition must be able to couple to zero. A photon carries one unit of angular momentum which is given up in the transition. In order for the angular momenta of the initial and final states and the photon to be able to couple to zero, the initial and final states may not differ in their angular momentum by more than 1 unit.

In single-electron atoms with only a single source of orbital angular momentum the resulting selection rule is

$$\Delta l = \pm 1. \tag{6.2}$$

When there are multiple electrons carrying orbital angular momentum, $\Delta L = 0$ is allowed as each electron is not necessarily precessing around the axis of the total orbital angular momentum vector, $\mathbf{L}$. This leads to a triangle selection rule, as the vector addition (tip to tail) to zero of the initial and final total orbital angular momentum vectors and the photon angular momentum vector resembles a triangle.

$$\Delta L = 0, \quad \pm 1 \tag{6.3}$$

However, $L = 0 \rightarrow L = 0$ is not permitted as it cannot satisfy a triangle condition.

Rule 3: Projection

Photons have a helicity of ± 1, for left- and right-circular polarisation, respectively (but not of zero because they are moving at the speed of light). If the photon and the orbital angular momentum of an electron are projected on to the same axis, then conservation of orbital angular momentum dictates that the m_l of the wavefunction must change by $+1$ on absorption of a left-circularly polarised photon and -1 on absorption of a right-circularly polarised photon. However, when the axes of quantisation are not the same, $\Delta m_l = 0$ is also permitted. Similar conclusions are reached for multi-electron states and for total angular momentum quantum numbers, when spin and orbital angular momenta have coupled to a resultant, leading to the following selection rules:

$$\Delta m_l = 0, \pm 1 \quad \Delta m_j = 0, \pm 1 \quad \Delta M_L = 0, \pm 1 \quad \Delta M_J = 0, \pm 1. \tag{6.4}$$

Rule 4: Spin

Even though the angular momentum of a photon is spin angular momentum, it only interacts with the orbital angular momentum of electrons. This is because the photon moves the electron through space when it interacts with it. The photon has no effect on electrons' intrinsic angular momentum, leading to the spin selection rules, which apply equally to single- and multiple-electron states, and to the projection as well as the total spin angular momentum.

$$\Delta s = 0 \quad \Delta S = 0 \quad \Delta m_s = 0 \quad \Delta M_S = 0 \tag{6.5}$$

Note that the spin–orbit interaction involves the coupling of the spin angular momentum to the orbital angular momentum to give a resultant total angular momentum; the photon can interact with the total angular momentum. Therefore the spin selection rule is relaxed as spin–orbit coupling becomes significant. Since the total angular momentum will also interact with an electric field, the spin–orbit interaction allows for manipulation of electron spin in 'spintronic' devices with electric fields [87].

Rule 5: Total angular momentum

When spin–orbit coupling is significant, the spin and orbital angular momentum of an electron couple to a resultant total angular momentum, j. On the grounds of conservation of angular momentum, Δj cannot exceed 1. Since the **l** and **s** vectors do not necessarily precess around a common axis, $\Delta j = 0$ is permissible. The same selection rule applies to multi-electron states.

$$\Delta j = 0, \pm 1 \quad \Delta J = 0, \pm 1 \tag{6.6}$$

Exercise 73

Some transitions are regarded as more forbidden than others. Suggest why the electric-dipole transition from the ground state to $^5D_0$ in Eu^{3+} ($f^6 \rightarrow f^6$) may be regarded as particularly forbidden.

Application to transition metal ions

Sixth-form students are often taught that transition-metal complex ions are coloured on account of d–d electronic transitions. It is therefore curious that these transitions are Laporte-forbidden, and that some of the most intensely coloured transition metal species have d^0 configurations, e.g. MnO_4^-. Mn^{7+} d^0 would be too polarising to exist in aqueous solution. In the MnO_4^- complex ion, the oxide ions are so polarised that there are effectively polar-covalent Mn–O bonds, so we may apply some of the ideas from our consideration of heteronuclear diatomic molecules. The electronegative oxygen atom has its valence electrons at a lower energy than the valence orbitals of manganese. Hence the polar-covalent bonding MO resembles the oxide AOs. When light of the appropriate frequency is absorbed by these bonding electrons, one is promoted to a higher energy orbital, which will resemble a manganese AO. Hence these transitions are known as ligand-to-metal-charge-transfer (LMCT) transitions. The reason they can be so intense is because they are not restricted by the Laporte selection rule: tetrahedral complex ions (like MnO_4^-, CrO_4^{2-}, etc.) have no centre of symmetry and so there can be no parity selection rule. Furthermore, the electron is being excited from an orbital of mainly p character to one of d character: being of opposite parity makes the transition more Laporte-allowed.

A similar problem is encountered with d^{10} transition metal compounds that are coloured, in that there is no possibility of d–d transitions. Indeed cadmium and mercury sulfides are used as pigments — cadmium yellow and vermilion, respectively. The polarisability of the sulfide and metal ions gives some covalent character to their bonding interaction. As a result they form semiconductors. The energy gap between the valence and conduction bands (bonding and antibonding orbitals, effectively) corresponds to visible light and so accounts for their colour.

The question then arises why d–d transitions are observed at all if they are forbidden by the Laporte selection rule. This selection rule only applies when there is a centre of symmetry. Vibrations of the metal-ligand bonds will remove the centre of symmetry if the group vibration is asymmetric, which relaxes the selection rule. Electronic transitions made viable by vibrations in this way are known as vibronic transitions. Such vibrations also lead to broadening of the spectral absorptions, as the vibration modulates the overlap of the ligand orbitals with the metal orbitals.

The anomalously pale Mn^{2+} d^5 ion was discussed in Section 3.10. Its d–d transitions are forbidden by the spin selection rule. However, it is not completely colourless but a pale pink, which raises the question of why it isn't completely colourless. Its colour can be explained by the spin–orbit interaction. The Mn^{2+} ground-state Russell–Saunders term is ^6S, with total angular momentum $J = \frac{5}{2}$. It is the only sextet term of d^5. The next highest spin multiplicity is quartet; the highest possible L quantum number for a d^5 quartet is 4. Hence the first excited Russell–Saunders term is ^4G, whose J value can take the values $\frac{11}{2}, \frac{9}{2}, \frac{7}{2}, \frac{5}{2}$. Two states with the same J value can interact through the spin–orbit interaction, described in Section 4.5. This mixes the states slightly, giving some quartet character to the ground state, which relaxes the spin selection rule. Another justification is that S ceases to be a good quantum number when there is spin–orbit coupling, making the spin selection rule redundant.

A similar argument can be made using crystal-field states (described in Section 4.9). The fully symmetric S term gives the crystal-field state A_{1g}. The first excited G term also gives A_{1g} as one of its crystal-field states. Two states with the same crystal-field label can interact through the crystal-field interaction. Again, this mixes the states slightly, giving some quartet character to the ground state, which relaxes the spin selection rule.

6.2 Simple Model Systems

Before considering the spectra of atoms we will build up to it with simpler model systems that should illustrate the important principles.

6.2.1 *Particle in a 1D box*

The simplest possible model for a constrained electron is to have one in a one-dimensional box. We assume there are no forces or potentials within the box, and infinite potential at the walls, i.e. infinitely high barriers keeping the electron within the box. As a result the wavefunction is 0 at the walls. We consider the wavefunction to be a standing one-dimensional wave with nodes at the walls, separated by the length of the box, L.

The wavefunction will need to be a solution to the one-dimensional Schrödinger equation for kinetic energy:

$$\frac{-\hbar^2}{2m}\frac{d^2\psi(x)}{dx^2} = E\psi(x). \tag{6.7}$$

In order for equation (6.7) to be an eigenvalue equation the wavefunction must return itself after taking its second derivative with respect to x. The sine function would therefore be suitable. The simplest possible sine function for the purpose of the 1D box would be one with nodes at the

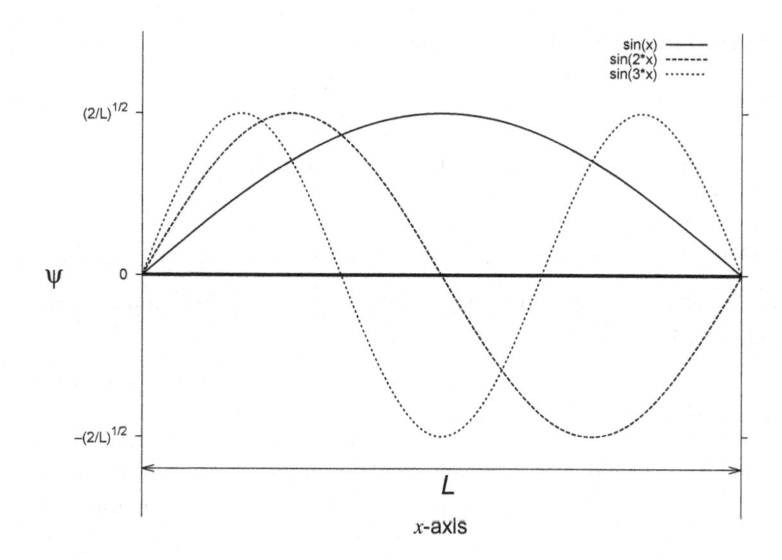

Figure 6.1. Wavefunctions in a one-dimensional box.

two walls but no nodes between. Such a function would be $\psi(x) = \sin\frac{\pi x}{L}$ as when $x = 0$ and L, $\psi(x) = 0$. Higher harmonics of this sine curve with nodes between the two walls will also be solutions to equation (6.7). Given that the operator for kinetic energy involves the second derivative of the wavefunction, the more tightly curved harmonics will represent higher energy states.

The necessity for the wavefunction to be zero at the walls of the box introduces the boundary condition that leads to the quantisation of the wavefunction, and therefore the quantum number n. The first three wavefunctions, with $n = 1, 2, 3$ are shown in Figure 6.1. Note that the ground state is $n = 1$, not $n = 0$ which would simply negate the wavefunction.

Another requirement for the wavefunction is a normalisation constant, N, to determine the amplitude of the wavefunction such that it may be properly normalised. A normalised wavefunction is one that, when multiplied by its complex conjugate, integrates over all space to give 1. This is tantamount to fixing the probability of finding the electron anywhere in all space being equal to 1. Integrating over 'all space' in the context of the 1D box means integrating between $x = 0$ and $x = L$. The 1D box wavefunction therefore has the form

$$\psi(x) = N \sin\left(\frac{n\pi x}{L}\right). \tag{6.8}$$

The value of N that normalises the wavefunction is $\sqrt{2/L}$.

Exercise 74

Show that the normalisation constant, N, for the wavefunction for a particle in a 1D box is $\sqrt{2/L}$.

Exercise 75

The eigenfunctions of a Hermitian operator should be mutually orthogonal. Show that the wavefunctions with $n = 1$ and $n = 2$ are orthogonal.

Exercise 76

Use the trigonometric identity below to show that all the wavefunctions for a particle in a 1D box are mutually orthogonal.

$$\sin(n_1 x)\sin(n_2 x) = \frac{1}{2}\left(\cos\{(n_1 - n_2)x\} - \cos\{(n_1 + n_2)x\}\right) \tag{6.9}$$

We now consider selection rules. Since the wavefunction has no orbital or spin angular momentum, parity symmetry is the only relevant selection rule. Considering the point half-way along the box to be the centre of symmetry, we can see from Figure 6.1 that the first three wavefunctions have parity symmetry of even, odd, even, respectively: in passing from the position $\frac{L}{2} - x'$ to $\frac{L}{2} + x'$ the sign of ψ stays the same for $n = 1$ and $n = 3$ and changes for $n = 2$. The sequence even, odd, even is what we see in the series of subshells s, p, d. Hence the transition $n = 1 \to n = 2$ should be allowed and $n = 1 \to n = 3$ should be forbidden by parity.

Exercise 77

Show that the transition $n = 1 \to n = 2$ is allowed and $n = 1 \to n = 3$ is forbidden by parity. Try to generalise to all different-parity and all same-parity transitions.

We can find the energy of the different quantum states in the 1D box and therefore the energy of transitions between them using the Schrödinger equation:

$$\begin{aligned}\frac{-\hbar^2}{2m}\frac{\mathrm{d}^2\psi(x)}{\mathrm{d}x^2} &= \frac{-\hbar^2}{2m}\frac{\mathrm{d}^2}{\mathrm{d}x^2}\left\{\sqrt{\frac{2}{L}}\sin\left(\frac{n\pi x}{L}\right)\right\} \\ &= \frac{\hbar^2}{2m}\sqrt{\frac{2}{L}}\left(\frac{n\pi}{L}\right)^2\sin\left(\frac{n\pi x}{L}\right) \\ &= E\sqrt{\frac{2}{L}}\sin\left(\frac{n\pi x}{L}\right). \end{aligned} \tag{6.10}$$

We can now write the energy of the quantum states as a function of the quantum number n. We use $\hbar = h/2\pi$.

$$E = \frac{h^2 n^2}{8mL^2} \tag{6.11}$$

It is clear from equations (6.10) that the value of the normalisation constant has no effect on the energy of the state. It can be seen that as the length of the box, L, is increased the energy of the wavefunction decreases. This follows from the reduction in the curvature of the wavefunction as the box length increases, which reduces the size of the second derivative of the wavefunction.

It is straightforward to find the energy gap between levels n and $n+1$:

$$\begin{aligned}\Delta E = E_{n+1} - E_n &= \frac{h^2}{8mL^2}\left\{(n+1)^2 - n^2\right\} \\ &= \frac{h^2}{8mL^2}(2n+1). \end{aligned} \tag{6.12}$$

Exercise 78

If a particle of mass 1 kg is moving at 1 m/s in an idealised 1D box of length 1 m, calculate its quantum number n and the energy required to excite it to the next quantum state. Comment on this value.

The 1D box model can be applied to straight-chain conjugated molecules where π electrons are delocalised along a section of the molecule which can be considered the box length, L. It is well known that as the extent of conjugation in molecules increases the wavelength of absorbed light also increases, i.e. the energy gap between ground and excited states in the molecule decreases with increasing box length.

Exercise 79

Find an expression for the number of conjugated C–C bonds, N, in a delocalised system for a molecule to absorb in the visible spectrum, and then solve it to find N. Take the blue end of the visible spectrum to be 400 nm, and the length of a C–C bond, l, to be 140 pm. Assume that each level denoted by a value of n can hold two electrons and that each carbon atom can add one electron to the π system. ($h = 6.6 \times 10^{-34}\,\text{Js}, m = 9.1 \times 10^{-31}\,\text{kg}, c = 3.0 \times 10^{8}\,\text{ms}^{-1}$.)

6.2.2 *Particle in a 2D box*

In the 2D box there is an additional degree of freedom, the y-axis, so that the particle exists on the xy-plane. Like the x-axis, in the y-direction, being bound between two infinitely high potentials a distance of L_y apart, a sine function of the same form of equation (6.8) can describe the wavefunction. In accordance with Section 2.2, the additional degree of freedom leads to a second quantum number. The wavefunction in equation (6.13) makes it clear how the x- and y-directions are described independently:

$$\psi(x, y) = \sqrt{\frac{2}{L_x}} \sin\left(\frac{n_x \pi x}{L_x}\right) \sqrt{\frac{2}{L_y}} \sin\left(\frac{n_y \pi y}{L_y}\right). \tag{6.13}$$

The wavefunctions are now described by two quantum numbers, n_x and n_y, that can vary independently of each other. Some examples of the wavefunctions, labelled with their two quantum numbers, are shown in Figure 6.2.

The energy of the wavefunction is therefore the sum of the energies in each direction. It will be given by the two-dimensional Schrödinger equation for kinetic energy. Now that there are two variables, the differential operators need to be partial differentials:

$$\frac{-\hbar^2}{2m}\left(\frac{\partial^2 \psi(x, y)}{\partial x^2} + \frac{\partial^2 \psi(x, y)}{\partial y^2}\right) = E_x \psi(x, y) + E_y \psi(x, y). \tag{6.14}$$

Each differential operator only acts on the half of the wavefunction containing the relevant variable, returning the same eigenvalue as was seen in equation (6.10). The half of the wavefunction containing the other

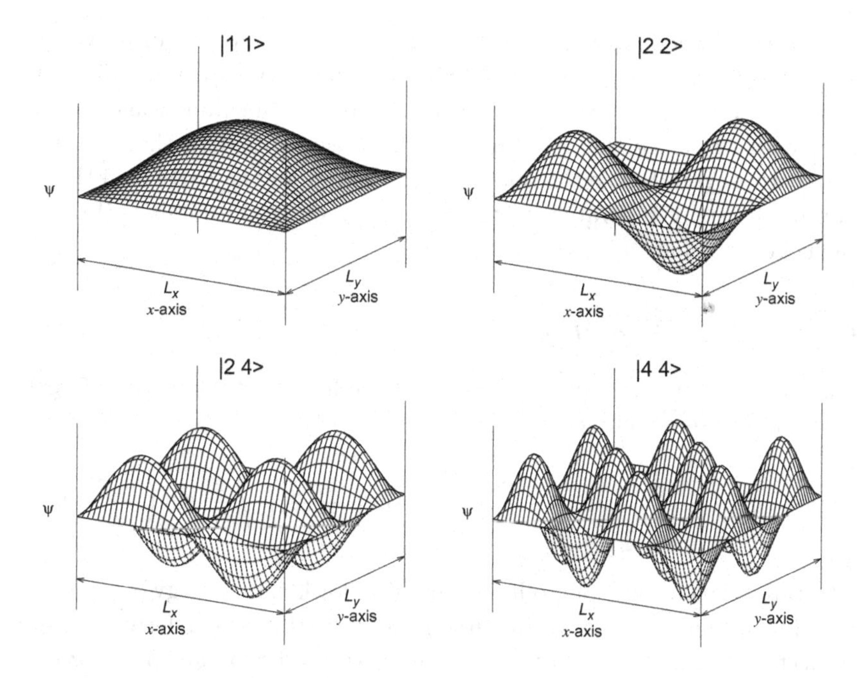

Figure 6.2. Wavefunctions in a two-dimensional box.

variable has no effect, merely cancelling out:

$$E_x = \frac{h^2 n_x^2}{8mL_x^2} \tag{6.15}$$

$$E_y = \frac{h^2 n_y^2}{8mL_y^2} \tag{6.16}$$

$$E_{\text{total}} = \frac{h^2}{8m}\left(\frac{n_x^2}{L_x^2} + \frac{n_y^2}{L_y^2}\right). \tag{6.17}$$

Exercise 80

Show that the square of the average value, $(<x>)^2$, is different from $< x^2 >$.

Exercise 81

Show that the average value of xy, $< xy >$, is the same as $< x >< y >$.

Electromagnetic radiation could be polarised along the x- or y-axis and so a dipole operator along either axis could interact with the half of the wavefunction on that axis. Each half of the wavefunction would behave independently since the other half would integrate to 1 in the transition moment integral. Hence the selection rules found for the particle in the 1D box would apply separately to the n_x and n_y quantum numbers if one of them were to change.

6.2.3 *Particle in a 3D box*

For a particle in a 3D box we can extend the findings for a 2D box to give the form of the wavefunction:

$$\psi(x,y,z) = \sqrt{\frac{2}{L_x}}\sin\left(\frac{n_x\pi x}{L_x}\right)\sqrt{\frac{2}{L_y}}\sin\left(\frac{n_y\pi y}{L_y}\right)\sqrt{\frac{2}{L_z}}\sin\left(\frac{n_z\pi y}{L_z}\right). \tag{6.18}$$

The three-dimensional Schrödinger equation for kinetic energy extends the two-dimensional one in equation (6.14) with a partial second derivative with respect to z. The energy of the particle likewise takes on a similar form:

$$E = \frac{h^2}{8m}\left(\frac{n_x^2}{L_x^2} + \frac{n_y^2}{L_y^2} + \frac{n_z^2}{L_z^2}\right). \tag{6.19}$$

Again, the selection rules obtained for the 1D box will apply independently to the three quantum numbers, n_x, n_y and n_z, if one of them were to change. If the three dimensions of the box, L_x, L_y and L_z, are equal, the box is cubic and the energy expression is simplified:

$$E = \frac{h^2}{8mL^2}(n_x^2 + n_y^2 + n_z^2). \tag{6.20}$$

$L = L_x = L_y = L_z$. In the cubic box, degeneracy arises in the excited states. Let us quote energy levels as multiples of $k = \frac{h^2}{8mL^2}$. The ground state is at $3k$.

Exercise 82

For a particle in a cubic box, work out how many excited energy levels there are within $12k$ of the ground state, where $k = \frac{h^2}{8mL^2}$. For each excited state give its energy in multiples of k, its degeneracy, g, and its parity.

Exercise 83

A colour centre in a crystal occurs when an electron becomes trapped in a vacancy caused by a missing ion. This may be modelled as a particle in a cubic box. A crystal of potassium chloride (normally colourless) appears magenta (blue mixed with red) because of colour centres in chloride vacancies. The electron absorbs green light at 510 nm, promoting it to the first excited state. Calculate the length of the cube in which the electron is trapped. What colour might a potassium bromide crystal be with this sort of colour centre? ($h = 6.63 \times 10^{-34}\,\mathrm{J\,s}, m = 9.11 \times 10^{-31}\,\mathrm{kg}, c = 3.00 \times 10^{8}\,\mathrm{m\,s^{-1}}$.)

6.2.4 *Particle on a ring*

We now consider a particle on a flat circular ring, with the potential energy being zero on the ring and infinite elsewhere. While a circle appears two-dimensional, if we have our centre of coordinates at its centre and use polar coordinates it becomes a one-dimensional problem in φ since r is fixed (as discussed in Section 2.2).

To solve the Schrödinger equation for kinetic energy we need to express the second derivative of the wavefunction with respect to x and y in terms of the polar coordinate φ. We can achieve this using the spherical polar analogues. The first derivatives of the spherical polar coordinates with respect to Cartesian coordinates are given in equations (A.3). If we ignore all terms containing z and θ we are left with the first derivatives of the polar coordinates with respect to Cartesian coordinates. The second derivatives of the spherical polar coordinates with respect to Cartesian coordinates are given in equations (A.5) and (A.9). These may similarly be converted to the second derivatives of polar coordinates with respect to Cartesian coordinates. Likewise we can convert the second derivative with respect to x (equation (A.8)) and with respect to y (equation (A.12)) of the wavefunction from the spherical polar calculations into their analogues for polar coordinates. These two spherical polar second derivatives are added to give the overall Laplacian in spherical polar coordinates.

There is a slight complication when converting the Laplacian in spherical polar coordinates of equation (A.13) to polar coordinates: the 2 in the spherical polar Laplacian arises from the sum of $\sin^2\varphi + \cos^2\varphi$ and $\sin^2\theta + \cos^2\theta$. Since we disregard terms in θ when converting spherical polar to polar coordinates, the 2 in equation (A.13) needs to be converted to a 1:

$$\nabla^2\psi = \frac{\partial^2\psi}{\partial r^2} + \frac{1}{r}\frac{\partial\psi}{\partial r} + \frac{1}{r^2}\frac{\partial^2\psi}{\partial\varphi^2}. \tag{6.21}$$

For the particle on a ring the Laplacian can be further simplified since, with r constant, we can ignore the differentials with respect to r:

$$\nabla^2\psi = \frac{1}{r^2}\frac{\partial^2\psi}{\partial\varphi^2}. \tag{6.22}$$

The Schrödinger equation for the kinetic energy of the particle on a ring is therefore

$$\frac{-\hbar^2}{2mr^2}\frac{\partial^2\psi(r,\varphi)}{\partial\varphi^2} = E\psi(r,\varphi). \tag{6.23}$$

It is convenient to rearrange this equation as follows:

$$\frac{\partial^2\psi(r,\varphi)}{\partial\varphi^2} = -\frac{2IE}{\hbar^2}\psi(r,\varphi), \tag{6.24}$$

where $I = mr^2$ is the moment of inertia. A wavefunction that satisfies this equation is

$$\psi(r,\varphi) = N\mathrm{e}^{\mathrm{i}m\varphi}, \tag{6.25}$$

where $m = \pm\left(\frac{2IE}{\hbar^2}\right)^{1/2}$. The boundary condition, which ensures that the wavefunction is single-valued, is $\psi(\varphi) = \psi(\varphi + 2\pi)$.

$$N\mathrm{e}^{\mathrm{i}m\varphi} = N\mathrm{e}^{\mathrm{i}m(\varphi+2\pi)} = N\mathrm{e}^{\mathrm{i}m\varphi}\mathrm{e}^{2\pi\mathrm{i}m} \tag{6.26}$$

Equation (6.26) is satisfied when $\mathrm{e}^{2\pi\mathrm{i}m} = 1$ which is true when $m = 0, \pm1, \pm2\ldots$. Therefore m is the quantum number for the particle on a ring, as was seen in equation (2.3). From the expression for m we determine the energy of these states:

$$E = \frac{\hbar^2 m^2}{2I} = \frac{h^2 m^2}{8\pi^2 M_e r^2} \tag{6.27}$$

$$\Delta E_{m\to m+1} = \frac{h^2}{8\pi^2 M_e r^2}(2m+1). \tag{6.28}$$

where M_e is the mass of the electron. The second form of the energy expression is written to emphasise its similarity to the energy in the 1D box, equation (6.11): both expressions involve a single quantum number squared on the numerator and the dimension of the system squared on the denominator. This isn't surprising considering that both are 1D models.

The ground state for the particle on the ring will be the singly degenerate $m = 0$, which is just the amplitude N. Note that quantum numbers relating to rotational motion and angular momentum may take a value of zero, as we saw for l and m in atoms, while quantum numbers relating to the boxes of different dimensions and the total energy in a hydrogen atom may not. Hence the ground state for angular momentum has zero energy while the different boxes and the hydrogen atoms have finite energy in their ground state, the so-called 'zero-point' energy. Zero energy is possible for the angular momentum ground states as they have a wavefunction with no curvature, while the boundary conditions in the different boxes and the hydrogen atom require curvature (the kinetic energy depends on curvature through the second-derivative operator).

The excited $\pm m$ states are doubly degenerate and complex, implying rotation around the ring in opposite directions. These complex functions may be made real or purely imaginary by taking linear combinations, as described and justified in Section 2.4:

$$\begin{aligned} |m\rangle + |m\rangle &= 2N\cos m\varphi \\ |m\rangle - |m\rangle &= 2\mathrm{i}N\sin m\varphi. \end{aligned} \tag{6.29}$$

Some examples of these functions, labelled by the linear combination of m values, are given in Figure 6.3. The normalisation constant for the trigonometric linear combinations was shown to be $\frac{1}{\sqrt{\pi}}$ in equations (2.10), (2.11) and (2.12).

The amplitude of the wavefunction, N, is also the normalisation constant which, according to equation (2.8) is $\frac{1}{\sqrt{2\pi}}$. Following the arguments for a 1D box, different $\sin m\varphi$ functions should all be mutually orthogonal. Since cosine functions are just sine functions shifted by $\pi/2$ radians, different $\cos m\varphi$ functions will also all be mutually orthogonal. Trigonometric arguments also show that the $\sin m\varphi$ and $\cos m\varphi$ functions are all mutually orthogonal.

Exercise 84

Show that the $\sin m\varphi$ and $\cos m\varphi$ functions are all mutually orthogonal. Use the following trigonometric identity:

$$\sin(m_1x)\cos(m_2x) = \frac{1}{2}\left(\sin\{(m_1+m_2)x\} + \sin\{(m_1-m_2)x\}\right). \tag{6.30}$$

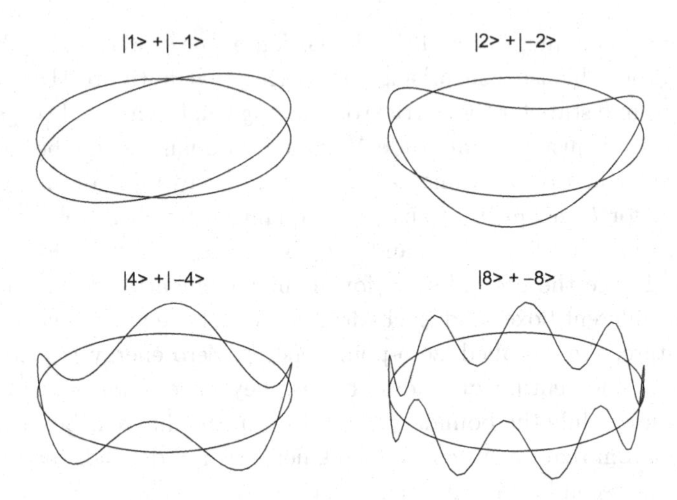

Figure 6.3. Wavefunctions on a ring.

As for parity symmetry, reflection of the wavefunction through the origin takes a point at φ to $\varphi+\pi$. For $\cos m\varphi$ and $\sin m\varphi$, $\mathrm{f}(\varphi) = \mathrm{f}(\varphi+\pi)$ when m is even, and $\mathrm{f}(\varphi) = -\mathrm{f}(\varphi+\pi)$ when m is odd. Hence odd m are odd parity and even m are even parity, which we can generalise as $(-1)^m$. This is consistent with the l labels of the atomic orbitals. Since the electric-dipole operator is odd parity, in order for the transition moment to be even parity and the transition therefore to be allowed, the two states being connected by the transition must be of different parity. This is consistent with the 1D box parity selection rule.

In addition to the parity selection rule there will also be an angular momentum selection rule since rotational motion is involved. This is easiest to establish when we consider the wavefunction in its complex form, $N\mathrm{e}^{\mathrm{i}m\varphi}$, and consider the electric-dipole operator to take the complex form $x \pm \mathrm{i}y$, since converting x and y in this complex to polar coordinates leads to $\mathrm{e}^{\pm\mathrm{i}\varphi}$. The transition moment for the transition $m_1 \rightarrow m_2$ is as follows:

$$\left\langle N\mathrm{e}^{\mathrm{i}m_1\varphi} \,\middle|\, \mathrm{e}^{\pm\mathrm{i}\varphi} \,\middle|\, N\mathrm{e}^{\mathrm{i}m_2\varphi}\right\rangle = N^2 \int_0^{2\pi} \mathrm{e}^{\mathrm{i}(-m_1\pm1+m_2)\varphi}\,\mathrm{d}\varphi. \tag{6.31}$$

The integral is only non-zero when $-m_1 \pm 1 + m_2 = 0$, which gives us the selection rule $\Delta m = \pm 1$.

There are mathematical parallels between the functions connected to a particle on a ring and those that describe the π MOs in benzene in

Section 5.6: both involve trigonometric functions derived from complex exponentials with a quantum number describing the number of nodal planes in the wavefunction. Indeed the benzene π system can be modelled using a particle on a ring.

Exercise 85

The most intense ultraviolet transition on benzene occurs at 184 nm, though there is one that is an order of magnitude weaker at 203.5 nm and a very weak forbidden vibronic transition around 254 nm [88]. (The vibronic transition is electronically forbidden but certain molecular vibrations allow the transition to be weakly observed by lowering the symmetry of the molecule.) Use the particle-on-a-ring model to predict the wavelength of the lowest energy ultraviolet absorption of the π electrons on benzene, and comment on the experimental data. Assume that each state can hold two electrons and that there are six π electrons. Take the circumference of the ring to be six times the C–C bond length in benzene of 140 pm (1 pm is 10^{-12} m). ($h = 6.626\times10^{-34}$ J s, $c = 3.00\times10^{8}$ m s^{-1}, $M_e = 9.11\times10^{-31}$ kg.)

6.2.5 *Particle on a circle*

The particle-on-a-circle model is relevant to the electron standing waves that have been observed in circular 'corals' of atoms adsorbed on surfaces [89]. It extends the one-dimensional (φ) particle-on-a-ring model to a second dimension, r, which will be free to vary from zero to the radius of the circle (the boundary condition). Following the discussion of Section 2.2, this additional degree of freedom should lead to an additional quantum number required to describe the eigenstates in this model.

Real and pure-imaginary linear combinations of the eigenfunctions for a particle in a circle give standing waves that are shown in Figure 6.4. They are labelled with their quantum numbers. The second quantum number is analogous to the m for a particle on a ring, taking the values 0, ±1, ±2, etc. The first quantum number is analogous to the radial quantum number in hydrogen, taking positive integer values, and with 1 for the ground state. As seen in the radial solutions for the hydrogen atom, each successive value for the radial quantum number is associated with an additional radial node. The signs in the figure give the phases of the wavefunctions.

The radial eigenfunctions for a particle on a circle are truncated Bessel functions, which are beyond the scope of this book. They are discussed in [90].

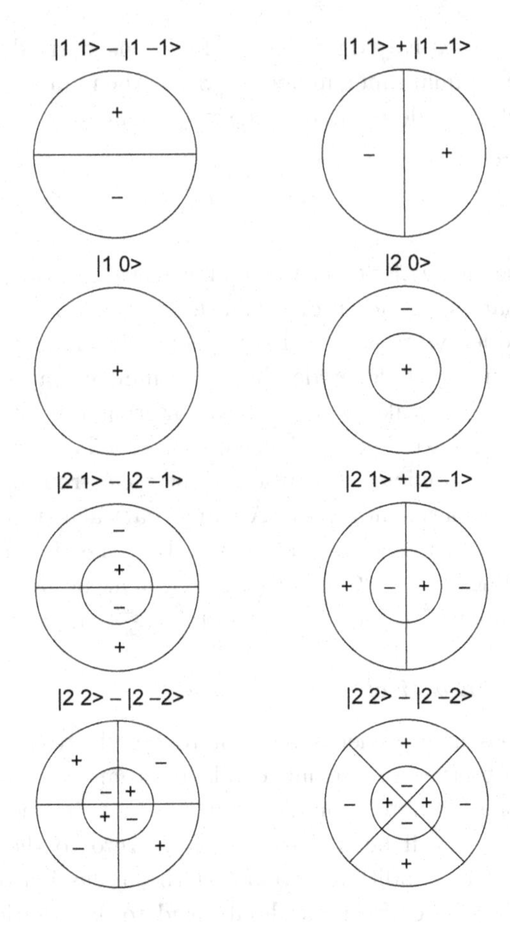

Figure 6.4. Wavefunctions in a circle. The signs show the phases.

6.2.6 *Particle on a sphere*

We assume that there is zero potential energy on the surface of the sphere and infinite potential energy everywhere else. While a sphere appears to be a three-dimensional object, points on a sphere can be plotted using just two variables, the angles θ and φ (the latitude and longitude, effectively), together with the constant r, the radius of the sphere. We therefore expect two quantum numbers to describe the eigenstates for the particle on the sphere.

The total energy of the particle is therefore its kinetic energy, $p^2/2m$, corresponding to the operator $-\hbar^2\nabla^2/2m$. ∇^2 is the Laplacian operator in spherical polar coordinates, shown in equation (2.18). The resulting expression can be simplified by neglecting the terms involving derivatives with respect to r. Alternatively, since the kinetic energy is purely rotational, we can use $l^2/2I$, using equation (A.14) to give the operator for l^2 in spherical polar coordinates. $I = mr^2$ is the moment of inertia of the particle of mass m. The result is equal to the Schrödinger equation for hydrogen, given in equation (2.22), neglecting the electron-nuclear electrostatic term and terms involving derivatives with respect to r:

$$\frac{-\hbar^2}{2mr^2}\left(\frac{1}{\sin\theta}\frac{\partial}{\partial\theta}\sin\theta\frac{\partial}{\partial\theta}+\frac{1}{\sin^2\theta}\frac{\partial^2}{\partial\varphi^2}\right)\psi = E\psi. \tag{6.32}$$

The solutions to this equation are therefore the spherical harmonic angular functions associated with the hydrogen atom, and so resemble the orbitals of Figures 2.2, 2.3, 2.4, 2.5 and 2.14. They are therefore specified by the quantum numbers, l and m_l.

We can therefore use the orbital representations to determine the parity selection rule as $(-1)^l$. Since the electric-dipole operator is odd-parity, it follows that the parity selection rules only allows transitions between states of opposite parity, i.e. connecting even and odd values of l.

The transition moment integral should give us the angular momentum selection rules. We first consider a z-polarised electric-dipole transition. We factorise the spherical harmonic eigenfunctions into an associated Legendre function of θ, $P_l^m(\theta)$ (collected in Table 2.1), and a complex exponential function of φ, $\mathrm{e}^{\mathrm{i}m\varphi}$. To integrate over both angles we use the Jacobian $\sin\theta\mathrm{d}\theta\mathrm{d}\varphi$.

It is relevant that the associated Legendre functions of rank l involve trigonometric functions raised to the power l (or the product of l such functions) and that different ranks are mutually orthogonal. Since $z = \cos\theta$, it can be equated to P_1^0. It follows that the l values associated with the two eigenfunctions in the integral may only differ by 1. Since multiplication of the three terms in the transition moment integral is associative, then the P_1^0 of the z-operator can be multiplied with the eigenfunction with l one smaller than the other eigenfunction. Then the integral will be over two associated Legendre functions of the same rank, which could give a non-zero result.

The z-operator is independent of φ, so the product of the two complex exponentials involving φ is just integrated between 0 and 2π. In order for this integral to be non-zero, $\Delta m = 0$.

For electric-dipole transitions polarised in the xy-plane, the process of finding the selection rules is analogous to the particle-on-a-ring treatment. This is because the operator is independent of θ and the associated Legendre functions are independent of φ. Hence the angular momentum selection rules for a particle on a sphere may be summarised as follows:

$$\Delta l = \pm 1 \quad \Delta m_l = 0, \pm 1.$$

Exercise 86

Use equation (6.32) and Table 2.1 to show that the energies of the $|lm\rangle$ eigenfunctions $|11\rangle$, $|10\rangle$ and $|1-1\rangle$ are independent of m.

Exercise 87

Use equation (6.32) and Table 2.1 to show that the energies of the $|lm\rangle$ eigenfunctions $|00\rangle$, $|10\rangle$ and $|20\rangle$ are equal to $l^2/2I$ where l^2 is given by the eigenvalue equation $\hat{l^2}\psi = \hbar^2 l(l+1)\psi$, which was quoted in Section 2.6.

We see from the exercises that the energies of the eigenfunctions are proportional to $l(l+1)$ and are $(2l+1)$-fold degenerate. We see that without the linear kinetic energy available when r is free to vary, the energy of the particle on a ring depends on l whereas the energy of the electron in a hydrogen atom does not.

Exercise 88

Show that the energy of a transition from l to $l+1$, in units of $\frac{\hbar^2}{2I}$ is $2(l+1)$, and that the energy gap between successive transitions is equal to 2 in the same units.

Exercise 89

Buckminsterfullerene, C_{60}, has the shape of a truncated icosahedron. Being the most symmetrical molecule in chemistry it has generated much discussion [91]. Its 60 π electrons might be modelled as particles on a sphere. Work out the orbital angular momentum quantum number, l, of its highest energy π electrons. Assume that each $|lm\rangle$ eigenfunction may associated with two electrons of opposite spin.

Exercise 90

Modelling the 60 π electrons of buckminsterfullerene, C_{60}, as particles on a sphere, the 10 most energetic electrons are among the 11 h orbitals. This would suggest C_{60} is paramagnetic (a consequence of unpaired electrons). In fact C_{60} and K_6C_{60} are diamagnetic (all electrons spin-paired) whereas K_3C_{60} is paramagnetic. Suggest how these observations may be rationalised using the fact that in an icosahedral crystal field the $l = 5$ level splits into three, with degeneracies of 5, 3 and 3.

6.3 Simple Atomic Spectra

Simple atomic spectra may be defined as those where the ground and excited states have no more than one electron with non-zero orbital angular momentum. Atoms with one, two or three valence electrons may therefore qualify as producing simple atomic spectra.

6.3.1 *Hydrogen*

The particle-on-a-sphere model is much like the hydrogen atom as far as angular momentum is concerned, and the same selection rules apply. The spin selection rule was given in Section 6.1.

Being a one-electron atom, its spectrum is far simpler than multi-electron atomic spectra; its most important features were essentially explained by Bohr in the first of his three landmark 1913 papers in *Philosophical Magazine* [92]. It was clear from Bohr's analysis and the original spectroscopic measurements by Balmer in 1885 that there is no restriction in the change in principal quantum number n in electronic transitions. Demonstrating this mathematically is more involved than was the case with the other selection rules. A more abstract method was developed by Dirac (from page 159 in [7]) and discussed by Sannigrahi and Das. It involves the commutation relations of constants of the motion and leads to the conclusion that Δn may be anything except 0 [93].

It should be noted that in these treatments and in the particle-on-a-sphere model, no consideration is given to magnetic and other relativistic effects. In this simplified approach we find the following selection rules:

$$\Delta n \neq 0 \quad \Delta l = \pm 1 \quad \Delta m_l = 0, \pm 1 \quad \Delta m_s = 0. \tag{6.33}$$

To a first approximation, the energies of electronic transitions on the hydrogen atom are given by the Rydberg equation, i.e. equation (2.34) with $Z = 1$. The Rydberg equation shows that at high values for the principal quantum number, n, the energy approaches the limit of $n = \infty$, where the electron is at the point of ionisation (and zero potential and kinetic energy, by definition).

The Rydberg equation also applies to other hydrogenic atoms with higher values of Z, such as He^+, Li^{2+}, etc., leading to qualitatively similar spectra. However, the larger value of Z found in other hydrogenic atoms means that the energy gaps between levels is larger.

Exercise 91

Calculate the energy of the hydrogen Lyman α-line ($n = 2 \rightarrow n = 1$) in terms of the Rydberg constant. Which transition in the spectrum of He^+ has the same energy?

All of the absorption transitions, i.e. electronic transitions in which the initial state is the ground state, are in the ultraviolet. In emission transitions from excited states the lines form series according to the level to which the electron drops. The names of the series and the final state of the electron in each, together with the limiting wavelength, are given in Table 6.1. The limiting wavelength corresponds to the transition from $n = \infty$. While the Lyman series is necessarily in the ultraviolet (like the absorption spectrum), the Balmer series is in the visible, and the remaining series are in the infrared.

The allowed electronic transitions are conveniently represented by a Grotrian diagram: the energy levels are plotted against angular momentum and allowed transitions are indicated with dotted lines. The diagram for hydrogen is given in Figure 6.5. As is conventional for spectroscopists, the energy is quoted in wavenumbers (reciprocal centimetres), related to energy through $E = hc\tilde{\nu}$, where h is Planck's constant, c is the speed of light in cm/s and $\tilde{\nu}$ is the quantity in wavenumbers (not, strictly, an energy of course).

The electronic transitions predicted by the Rydberg equation are split to a tiny extent by spin–orbit coupling. It is only observed at high resolution

Table 6.1. The series names and limits in the emission spectrum of the hydrogen atom.

Lower n	Name of series	λ_∞/nm
1	Lyman	91.2
2	Balmer	365
3	Paschen	820
4	Brackett	1458
5	Pfund	2280
6	Humphreys	3280

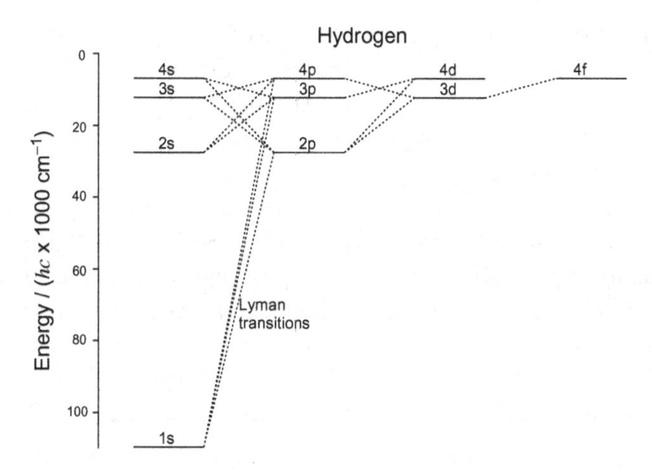

Figure 6.5. Grotrian diagram for hydrogen. Zero energy is defined in the diagram as the ionisation limit, $n = \infty$. Data taken from [31].

as the splittings are smaller than a wavenumber. Since the spin–orbit interaction energy depends on Z^4 [10] it is very small in hydrogen compared with most other elements. The ordering of spin–orbit coupled energies depends only on j, not the parent value of l or m_s; within a principal shell the energies increase with j. This was explained by Dirac who showed that the effect of combining the spin–orbit energy with the relativistic correction energy is to remove the dependence on s and l from the energy of the hydrogen electron. Equation (6.34), which he obtained, shows that the energy of an electron in the hydrogen atom depends only on n and j (page 119 of [63]):

$$E_{n,j} = -\frac{Z^2 R_H}{n^2}\left[1 + \frac{\alpha^2 Z^2}{n}\left(\frac{1}{j+\frac{1}{2}} - \frac{3}{4n}\right)\right], \tag{6.34}$$

where α is the fine structure constant.

The exception to the rule of j levels being degenerate within a principal shell is $s_{\frac{1}{2}}$, which is at a higher energy than $p_{\frac{1}{2}}$ (the subscript $\frac{1}{2}$ is the j value) in each shell — except $n = 1$ where there is no p subshell. The s subshell is exceptional because the Lamb shift is only appreciable for s electrons. The Lamb shift is caused by fluctuations in the electromagnetic vacuum which are most significant very close to the nucleus. Since electrons other than s electrons have a virtually zero probability of being within or extremely close to the nucleus, the Lamb shift only appreciably affects the s electrons. The Lamb shift has been measured at $0.0353\,\mathrm{cm}^{-1}$ for $2s_{\frac{1}{2}}$ and

$0.0105\,\text{cm}^{-1}$ for $3\text{s}_{\frac{1}{2}}$ (page 122 of [63]). Indeed the $2\text{s}_{\frac{1}{2}}$ level is metastable: the electric-dipole transition down to 1s is forbidden on the grounds of angular momentum conservation and the parity selection rule, while the transition to $2\text{p}_{\frac{1}{2}}$ is vanishingly weak due to the very small energy gap between the two levels.

The other interaction which can split the energy levels in the hydrogen atom is the hyperfine interaction, explained in Section 4.8. The spin of the nucleus in hydrogen, given by nuclear spin quantum number $I = \frac{1}{2}$ will couple with all of the half-integral spin–orbit j levels on hydrogen to give hyperfine levels described by the quantum number $F = j \pm \frac{1}{2}$, with the lower value of F being associated with a lower energy. This energy gap has been measured as $0.0474\,\text{cm}^{-1}$ for the 1s electron in hydrogen [31].

6.3.2 *Helium*

The electronic transitions on a helium atom fall into two mutually exclusive groups — singlet and triplet — as is shown in the Grotrian diagram in Figure 6.6. Indeed it was once thought that the spectral lines belonged to two elements. The spin selection rule prevents all but the weakest transitions between singlet and triplet states, as the very weak spin–orbit interaction

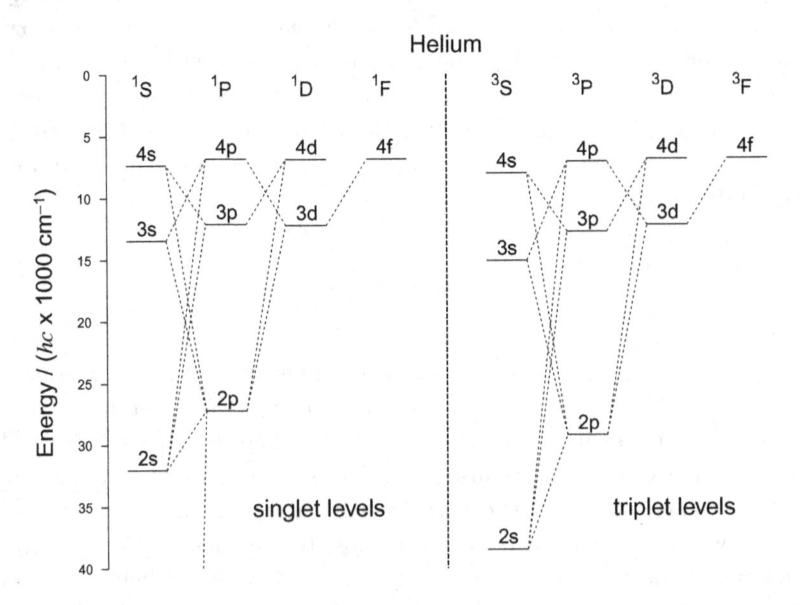

Figure 6.6. Grotrian diagram for helium. Zero energy is defined in the diagram as the ionisation limit, $n = \infty$. Data taken from [31].

for such a light element only relaxes the selection rule to a very small degree. Indeed the 1s2s ^{3}S state, lacking an allowed electric-dipole transition to the singlet ground state, is metastable. The 1s2s ^{1}S state is also metastable: the electric-dipole transition to the ground state is not forbidden by the spin selection rule, but is forbidden on angular momentum and parity grounds.

High resolution is required to observe the splitting of triplet peaks into three lines ($j = l + 1, l, l - 1$) as the splitting is typically much smaller than a wavenumber. Following Hund's rules (Section 4.6), the higher-j spin–orbit levels within a multiplet are lower in energy. ^{3}S states are not split at all as there is no orbital angular momentum to couple with the spin angular momentum. If the excited helium atoms are in equilibrium with one another then the intensity of the j components of the triplet levels should be proportional to their degeneracy, $2j + 1$. However, this is often not observed when working with low pressures of gases, where the atoms are quite isolated from one another.

Since the wavefunctions are for multi-electron states, term symbols are quoted above groups of similar terms in the Grotrian diagram. The wavefunctions that describe singlet and triplet states were described in Section 4.1. Following Hund's rules, the triplet levels are lower in energy than the analogous singlet levels. Since the triplet term is symmetric with respect to exchange of the two electrons' spins, the spatial one-electron wavefunctions must form a Slater determinant of the form $\psi_1(r_1)\psi_2(r_2) - \psi_1(r_2)\psi_2(r_1)$ so that they are antisymmetric with respect to inversion (Section 4.1). The Slater determinant vanishes when $r_1 = r_2$ and, as a result, the two electrons are less likely to be found very close together, producing a so-called Fermi hole in the probability density of the relative positions of the electrons. The extra distance between the two electrons is consistent with the lower energy of triplet terms on account of their exchange energies.

To a first approximation, the photon absorbed or emitted from an atom interacts with only a single electron. Hence the Grotrian diagram indicates the orbital occupied by the excited electron (whether as part of a singlet or a triplet state), with the other electron assumed to remain in the 1s orbital. As a result the atomic spectrum resembles somewhat two spectra of the alkali-metal type, except the lowest S term in the triplet spectrum appears missing (as 1s^2 cannot form a triplet term due to the Pauli principle). The ground-state energy is omitted from the Grotrian diagram as it is so distant from the excited states, at 198305 cm^{-1} or 24.6 eV. Indeed, in helium the energy of the first excited state is over 80% of the ionisation energy, while the figure for hydrogen is 75% [31]. This is because the effective nuclear charge for the more radially extended excited electron in helium will be lower than it is in the ground state.

Exercise 92

Assume that the electron that gets excited in a transition on helium is perfectly shielded from the nucleus by the other electron. Use the Rydberg equation (2.34) to determine the effective n value, n^*, for the ground state and the first excited state in absorption (1s2p ^{1}P), which have energies of 198305 cm^{-1} and 27176 cm^{-1}, respectively. $R_H = 109679\,\text{cm}^{-1}$. Comment on the values obtained.

Within a principal shell, the excited electron's energy depends on its orbital angular momentum (which was not the case in the hydrogen atom) due to the different interaction of the subshells with the other electron in the 1s subshell. Since the subshells within a principal shell are no longer degenerate, Δn for transitions of the excited electron is unrestricted.

$$\Delta n \text{ unrestricted} \quad \Delta l = \pm 1 \quad \Delta m_l = 0, \pm 1 \quad \Delta m_s = 0. \tag{6.35}$$

The intensity of transitions diminishes as the energy difference between the two states diminishes. Allowed transitions are not indicated in the Grotrian diagram between states of very similar energy — even when they are not formally forbidden — since their intensity will be very weak.

As was the case with the hydrogen atomic spectrum, species isoelectronic with helium, e.g. Li^+, Be^{2+}, etc., give qualitatively similar spectra to helium except the larger nuclear charge leads to larger energy differences between levels. An interesting exception is the H^- ion. The H atom may capture an electron and exist in the 1s^2 configuration since each electron experiences poor shielding from the nucleus by the other electron. However, the 1s2s excited state is not stable since the 1s electron largely shields the 2s electron from the nuclear charge, and it ionises spontaneously.

6.3.3 *Sodium*

The sodium atom has 11 electrons but 10 of them are in closed shells (1s^2 2s^2 2p^6). Summing the m_l and m_s values for all the closed-shell electrons gives zero for both M_L and M_S, giving them ^{1}S character. The closed-shell electrons therefore behave as a spherically symmetrical region of electron density shielding the 3s valence electron from the nucleus. The atomic spectrum of sodium might therefore be assumed to be hydrogen-like in appearance. However, like helium, the excited subshells that the valence electron might occupy are shielded differently by the remaining electron density. Hence the different subshells within a principal shell are of different energy, with the s subshell being the lowest in energy. Unlike helium, the excited states are not divided into singlets and triplets; rather, they are all doublets.

Alkali metal atomic spectra are important historically as they gave rise to the letters we now associate with the subshells — they are the simplest atomic spectra where the subshells are clearly distinct. Transitions to s subshells (from p subshells) were identified as sharp; transitions to p subshells (from s subshells) as principal; transitions to d subshells (from p subshells) as diffuse; and transitions to f subshells (from d subshells) as fundamental [63].

The Grotrian diagram for sodium is given in Figure 6.7. Some transitions are not indicated in the diagram despite not being formally forbidden, such as those between 4p and 5s, 4s and 3d. This is because they are very weak on account of the small energy gap between the levels. The lines labelled Fraunhofer D1 and D2 are by far the most intense of the emission spectrum, at around 589 nm, observed in the optical spectrum of the Sun by Fraunhofer in 1814.

The two Fraunhofer D lines account for the distinctive yellow colour of street lights. (They do not account for the colour of the Sun; the Fraunhofer lines are actually absences from the solar spectrum caused by

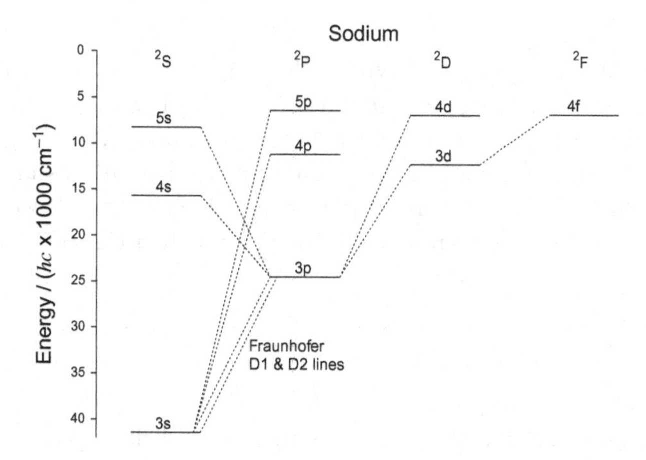

Figure 6.7. Grotrian diagram for sodium. Zero energy is defined in the diagram as the ionisation limit, $n = \infty$. Data taken from [31].

absorption of sunlight by sodium atoms.) Sodium is an ideal element for street lighting. The low ionisation energy of sodium means that most of the atomic spectrum is in the visible region. Since sodium vapour is mainly monatomic, the atomic spectrum is easy to produce.

Following the selection rules that applied to helium, we would anticipate that the sodium absorption spectrum was a series of 3s→ np lines ($n \geq 3$). All these absorption lines are doublets, like the Fraunhofer D lines. This is because the spin–orbit coupling of the electron spin ($s = \frac{1}{2}$) with the orbital angular momentum ($l = 1$) to give two levels with total angular momentum quantum number $j = l \pm s$.

Being a fairly light element, the splitting of the p subshell energies is not large enough to be seen at low resolution (in the limit of one-electron atoms, the spin–orbit interaction energy increases with Z^4 [10]). The splitting of the 3p doublet is $17.2\,\text{cm}^{-1}$; splittings are smaller for higher p subshells ($5.6\,\text{cm}^{-1}$ for 4p, $2.5\,\text{cm}^{-1}$ for 5p, $1.25\,\text{cm}^{-1}$ for 6p). In emission, all the lines are doublets except those to S states with $l = 0$. The splittings of the d subshells are all smaller than a wavenumber [31]. Following Hund's rules (Section 4.6), the lower j value in the doublet takes the lower energy.

The selection rules for sodium should also include the total angular momentum quantum number, j, and its projection on the z-axis, m_j. Following the discussion in Section 6.1 the following selection rules apply:

$$\Delta l = \pm 1 \quad \Delta m_l = 0, \pm 1 \quad \Delta m_s = 0 \quad \Delta j = 0, \pm 1 \quad \Delta m_j = 0, \pm 1. \tag{6.36}$$

Historically a quantity known as the quantum defect, δ_{nl}, has been associated with the energies of nl subshells in alkali metal atoms. It is the difference between the actual principal quantum number, n, of the valence electron and its effective principal quantum number, n^*, assuming that $Z_{\text{eff}} = 1$ [94]. It is a measure of the penetration of the valence electron into the inner core of electrons and so is principally a function of l rather than n.

Exercise 93

Use the following energies, measured from the ground state, to determine the quantum defect for the 3s, 3p, 3d and 4s subshells. Comment on the result. The ionisation energy of the sodium atom is $41450\,\text{cm}^{-1}$. Energies in wavenumbers: 3p 16968, 3d 29173, 4s 25740 [31]. $R_H = 109679\,\text{cm}^{-1}$.

6.3.4 *Groups 2 and 12*

The atomic spectra of elements in groups 2 (s^2) and 12 ($d^{10}s^2$) might be considered helium-like (since the d subshell in group 12 is closed). There are two changes to the helium-like appearance that emerge with increasing atomic number. Firstly, the increasing number of inner shielding electrons leads to a greater difference in the energies of subshells within a principal shell. Secondly, with increasing atomic number the spin–orbit interaction becomes more significant. With the increasing interaction between spin and orbital angular momentum, S and L become less good quantum numbers and transitions between singlet and triplet levels become less forbidden. Furthermore, as spin–orbit splittings become clearly resolved the spectra appear far more complicated.

Mercury lamps are often used to provide reference spectra when doing spectroscopy. With its large atomic number, spin–orbit splittings are clearly resolved and many singlet–triplet transitions are visible, leading to a great many transitions, making it a convenient reference source across a wide range of visible and ultraviolet frequencies. Furthermore, the volatility of mercury makes it easy to generate the atomic emission spectrum in a lamp. Mercury vapour is essentially monatomic, making it easy to produce the atomic emission spectrum.

The radial density function of 2p electrons in beryllium has been shown to be significantly contracted in triplet states compared to singlets [63]. This might be taken to be a result of the lowering of the triplet energies due to the exchange interaction.

6.3.5 *Group 13*

Despite having three valence electrons, group 13 elements produce simple spectra because the two valence s electrons constitute a closed subshell. The atomic spectra of group 13 elements (s^2p^1) are somewhat like alkali metal spectra, except the ground state of the valence electron is in a p subshell rather than an s. The spectra are somewhat more complex than those of alkali metals in absorption as from the valence p subshell there are allowed transitions to both s and d subshells. The emission spectra look like those of alkali metals, except the lowest s level is not seen.

The valence s^2 electrons are not particularly core-like, being in the same principal shell as the valence p electron. This leads to a greater difference in the energy of subshells within a principal shell than is the case with the alkali metals.

6.4 Complex Atomic Spectra

Atoms with two or more electrons with non-zero orbital angular momentum produce complex spectra. This is because the coupling of the orbital angular momentum on multiple electrons leads to multiple terms within the configuration. In simple spectra, by contrast, in the limit of no spin–orbit coupling each configuration consists of only a single energy level when there is just one electron outside closed subshells, and there is only a singlet and a triplet term for ss, sp, sd and sf configurations.

One would expect carbon, $2s^2\ 2p^2$, to be the first element to qualify for complex atomic spectra. However, as long as an element has more than one electron it is possible to generate complex spectra through two-electron excitations. In the orbital approximation, electrons are taken to occupy hydrogen-like orbitals (in the form of suitably antisymmetrised wavefunctions) and photons are assumed to act on single electrons at a time. Two-electron excitations signify a contravention of the orbital approximation.

6.4.1 *Displaced terms and closed shells*

Two-electron excitations lead to the displacement of series limits in spectra. This is evident in Figure 6.8 which compares one- and two-electron

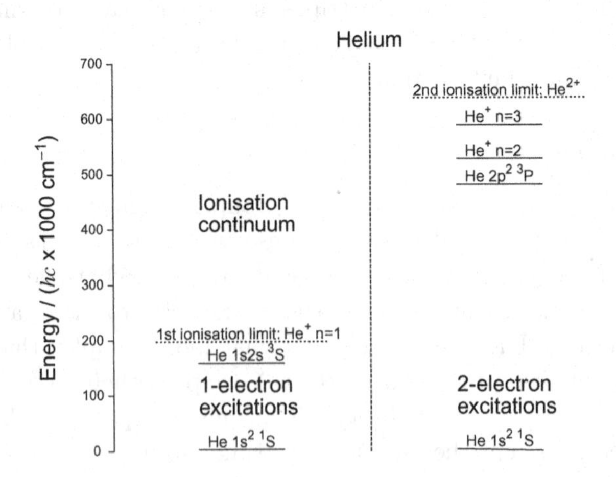

Figure 6.8. Simplified energy-level diagram for helium comparing one- and two-electron excitations. Zero energy is defined in the diagram as the $1s^2$ ground state for the helium atom. Data taken from [31].

excitations in helium. On the left-hand side the ionisation limit is the same as the one in the helium Grotrian diagram of Figure 6.6. Since just one electron is being excited, there is the remaining 1s electron in every configuration. When the excited electron reaches the ionisation limit, the remaining He^+ ion is therefore in its 1s ground state. Above the ionisation energy is a continuum of energy as the ejected electron, not being in a bound state, is effectively unquantised.

The right-hand side of the diagram shows that it is possible for the helium atom to exist in a bound state above its ionisation energy when both electrons are excited. $2p^2$ is the example given in the figure, which is a configuration that yields complex spectra. However, such states are short-lived as they are prone to autoionise into a continuum state. There will be series of doubly-excited configurations where one electron is in the second shell and the other is excited to successively higher shells. Such a series will converge to an ionisation limit, leaving the remaining He^+ ion in the excited $n = 2$ state (subshells will be of the same energy as He^+ is hydrogenic), indicated in the figure. This is an example of displaced terms. Likewise, there will be series of terms with one electron in the third shell, converging on the He^+ ion in the $n = 3$ excited state.

The group 2 elements, having an s^2 configuration like helium, would similarly be expected to form complex spectra following two-electron excitations. Calcium and the group 2 elements below it have even more complicated spectra because $((n-1))$d orbitals are close in energy to the ns valence orbitals; both $4p^2$ and $3d^2$ configurations are observed. This allows for many more transitions to be seen in the spectrum.

It is also possible to obtain complex from atoms with no net angular momentum at all. Noble gases have only closed shells; however, on exciting an electron from a full p subshell to a vacant subshell with non-zero angular momentum, two sources of orbital angular momentum result: the excited electron and the hole created in the p subshell. As was discussed in Section 4.6, a p^5 subshell behaves as a single positive electron, which can therefore couple of the excited electron.

As well as exciting an electron from the closed valence shell of a noble gas, complex spectra may also be generated by exciting an electron from a closed inner shell to a vacant shell. These have been studied for the heavy alkali metals. There are series in caesium, for example, from the excitation of a 5p electron; the series converge to the limit of Cs^+ [Kr]$4d^{10}$ $5s^2$ $5p^5$ 6s [63].

6.4.2 *Carbon*

The ground configuration of carbon, $2s^2\ 2p^2$, has two electron with non-zero orbital angular momentum and so will form a number of terms. These were discussed in Section 4.3. With multiple terms arising from each configuration, many more lines are observed in the spectra, including low energy transitions within the ground configuration, and the Grotrian diagram becomes rather crowded. Transitions within the ground configurations may however be very weak as they cannot involve a change in parity, and so are parity-forbidden. The atomic spectrum of carbon is complicated by the fact that terms within different configurations overlap with each other in energy.

The simplest transitions in carbon are those involving excitations from 2p to ns, where $n > 2$, since only one electron has non-zero orbital angular momentum in the excited state. These transitions satisfy the parity selection rule and are quite intense. All levels in this series are P states, and will be singlets or triplets, depending on the relative orientations of the spin of the unexcited 2p and the excited ns electron. The series converges on the ground state of C^+, $2s^2\ 2p$, ie the boron configuration. Displaced terms result from the excitation of a 2s electron to np levels; this series converges on the excited $2s\ 2p^2$ state of C^+.

A configuration of particular interest in carbon is $2s\ 2p^3$ as it is associated with sp^3-hybridised bonding. The terms associated with this configuration will be closely related to the terms of p^3 that were considered in Section 4.3 — ^2D, ^2P and ^4S. Since the additional s electron carries no orbital angular momentum there is no change in the L symbols, but the multiplicity of each term becomes either one greater or one lesser on account of the spin of the s electron either adding or subtracting from the resultant p^3 spin. Therefore the terms of sp^3 are ^3D, ^1D, ^3P, ^1P, ^5S and ^3S. According to Hund's rules (Section 4.6) the lowest energy term will be the highest spin-multiplicity ^5S, which has been observed at 33735 cm^{-1} [31]. The large range in spin multiplicities from quintet to singlet over the configuration results in a very large span of energies for the terms. Indeed the sp^3 singlet terms are higher in energy than the first ionisation limit forming C^+ $(2s^2\ 2p)$ at 90878 cm^{-1} [31].

Exercise 94

There is a parity-allowed $2p \rightarrow 3d$ transition from the ground state on carbon. Work out the Russell–Saunders terms in the excited configuration and suggest which is the lowest energy term.

For complex spectra, the selection rules for sodium, which were effectively one-electron selection rules, including spin–orbit coupling, need to be extended to the capital-letter quantum numbers for many-electron systems. They map across in predictable fashion, except the $\Delta l = 0$ selection rule becomes $\Delta L = 0, \pm 1$ as discussed in Section 6.1. The capital-letter selection rules apply to the Russell–Saunders terms. These, however, are relaxed by the spin–orbit coupling interaction.

$$\Delta l = \pm 1 \quad \Delta L = 0, \pm 1 \quad \Delta S = 0 \quad \Delta J = 0, \pm 1 \quad \Delta M_J = 0, \pm 1 \tag{6.37}$$

6.4.3 *Calculation of term energies for carbon*

The theory of Slater parameters can be applied to terms within a configuration in complex spectra to model the electrostatic interactions between the electrons. Knowledge of Slater parameters from observed term energies can be used to predict the energies of other terms in the configuration.

While the angular momentum characteristics of orbitals are well defined — even for multi-electron atoms in the orbital approximation — their radial characteristics are much more difficult to calculate for multi-electron atoms as they will depend sensitively on all the many-body interactions between the electrons. The approach therefore is to factorise the wavefunctions into radial and angular parts and to treat the radial part as a parameter to be fitted to the observed energies of the terms. There are typically many more terms than parameters, so the parameters can usually be reliably fitted to the term energies.

The electrostatic energy between two electrons is proportional to e^2/r_{12} where e is the charge of an electron and r_{12} is the distance between the two electrons. Using the spherical harmonic addition theorem it is possible to expand this expression out into a series of products of two spherical harmonic functions, one for each electron, where one spherical harmonic function is the complex conjugate of the other [28]. This is a convenient form for an operator as the angular wavefunctions of the electrons can also be expressed as a series of pairs of spherical harmonic functions:

$$\sum_{l=0}^{\infty} \sum_{m=-l}^{+l} Y_l^m (Y_l^m)^*. \tag{6.38}$$

Ultimately, to find the electrostatic interaction energies of a pair of electrons in a configuration, a Hamiltonian matrix needs to be assembled with the rows and columns each representing the composite states (or basis) of the configuration, and each matrix element at row r and column c being the integral $\left\langle \psi_r \middle| \hat{H} \middle| \psi_c \right\rangle$. Such a matrix needs to be diagonalised to give the energies of the pure states, as these lie along the main diagonal. This is effectively what was done with the secular determinants of Sections 5.2 and 5.6.

It is difficult to proceed with the Russell–Saunders terms as basis states of the configuration as these multi-electron states are described with determinantal functions. The best way to proceed is using the M_L and M_S quantum numbers to label the uncoupled microstates. The energy of a Russell–Saunders term, ${}^{2S+1}L$, does not depend on the value taken by the projections M_S and M_L; that is to say that the operator for electrostatic energy commutes with the M_S and M_L angular momenta. (Commutation was discussed in Section 1.6.) Therefore a Hamiltonian matrix for the electrostatic interaction between two electrons constructed in terms of M_S and M_L quantum numbers will only consist of diagonal terms, simplifying the calculation considerably.

If an $M_S M_L$ state corresponds to a single microstate of two electrons, we can find the electrostatic energy of interaction between the electrons using the pair of spherical harmonic functions as follows. The integral

$$\left\langle \psi(M_S, M_L) \middle| \sum_{l=0}^{\infty} \sum_{m=-l}^{+l} Y_l^m (Y_l^m)^* \middle| \psi(M_S, M_L) \right\rangle$$

may be factorised into two integrals, one for each electron in the two-electron microstate, each being operated on by one of the spherical harmonic operators, to give the angular part of the electrostatic interaction energy between the electrons. The spin of the electron doesn't affect the electrostatic calculation so from now on it will be omitted from these integrals and the m_l quantum number will be abbreviated to m. In order for the integrals to be non-zero, the sum of the three m terms must come to zero. The only m value of the spherical harmonic operators that will give a non-zero result is $m = m_{\text{bra}} - m_{\text{ket}}$.

$$\sum_{l=0}^{\infty} \left\langle \psi_{l_1 m_1} \middle| Y_l^0 \middle| \psi_{l_1 m_1} \right\rangle \left\langle \psi_{l_2 m_2} \middle| Y_l^0 \middle| \psi_{l_2 m_2} \right\rangle \propto E_{M_S M_L}(\text{Coul.}) \qquad (6.39)$$

Since wavefunctions are antisymmetric with respect to electron exchange, we need to subtract the integrals with the electrons exchanged. This negative term is the angular part of the exchange interaction energy between the electrons. Now the spin of the electrons is vital: if the two electrons are spin antiparallel then the integrals vanish.

$$\delta(m_s(1), m_s(2)) \sum_{l=0}^{\infty} \left(\langle \psi_{l_1 m_1} \left| Y_l^{m_1 - m_2} \right| \psi_{l_2 m_2} \rangle\right)^2 \propto E_{M_S M_L}(\text{Exch.}) \tag{6.40}$$

The δ term is a Kronecker delta. $\delta(m_s(1), m_s(2)) = 1$ when $m_s(1) = m_s(2)$; $\delta(m_s(1), m_s(2)) = 0$ when $m_s(1) \neq m_s(2)$.

The integrals of equations (6.39) and (6.40) are often written in the shorthand form $c^k(l_1 m_1, l_2 m_2)$, where k is the rank of the spherical harmonic operator, and the quantum numbers of the two electrons are given in brackets.

Now that the m values of the spherical harmonic operators have been determined, we turn to their l values, i.e. their rank, k. Spherical harmonic functions are associated with a parity (symmetry with respect to inversion through the origin) just as orbital subshells are, being real linear combinations of spherical harmonic functions. The product of the spherical harmonic functions of the bra, ket and operator must be even-parity if it is to integrate over all space to a non-zero result, giving the following selection rule:

$$k + l_1 + l_2 = \text{even}. \tag{6.41}$$

Furthermore, the triangle condition must be satisfied so that the integral gives a scalar result:

$$|l_1 - l_2| \leq k \leq l_1 + l_2. \tag{6.42}$$

All the spherical harmonic integrals can be solved analytically to give the angular contribution to the electrostatic interaction energy of the two electrons. The radial contribution to each integral is collected into Slater F^k parameters for direct-coulombic integrals between like electrons, and G^k parameters for exchange-coulombic integrals involving electrons from different subshells. When both electrons are from the same subshell, F^k parameters can also be used for the exchange integrals. The values taken for the F^k and G^k parameters for a given atom will depend on the particular subshells being described.

When an $M_S M_L$ state corresponds to a single uncoupled microstate of two electrons, we can express the total direct-coulombic interaction energy, J_{12}, between the two electrons, labelled 1 and 2, in terms of the

Table 6.2. The value of the $c^k(m_1, m_2)$ integrals for two p electrons [28].

m_1	m_2	c^0	$5c^2$
± 1	± 1	1	-1
± 1	0	0	$+\sqrt{3}$
0	0	1	$+2$
± 1	∓ 1	0	$-\sqrt{6}$

c^k expressions above:

$$J_{12} = \sum_{k=0}^{\infty} c^k(l_1m_1, l_1m_1)c^k(l_2m_2, l_2m_2)F^k. \tag{6.43}$$

Similarly, we can express the total exchange-coulombic interaction energy, K_{12}, between the two electrons, 1 and 2, in terms of the c^k:

$$K_{12} = \sum_{k=0}^{\infty} (c^k(l_1m_1, l_2m_2))^2 G^k. \tag{6.44}$$

The calculation of these angular integrals is beyond the scope of this book. In practice they tend to be obtained through their relationship with 3-j symbols, which are mathematically convenient objects that are simple to compute. Interested readers should consult Rotenberg [95]. The c^k integrals are given for two p electrons (i.e. $l = 1$) in Table 6.2 [28]. The k ranks are limited to 0 and 2, which conform to both parity and triangle selection rules.

The fully symmetric $c^0(m_1, m_2)$ integrals are always equal to 1 for direct-coulombic integrals, i.e. when $m_1 = m_2$, and always equal to zero for exchange-coulombic integrals, i.e. when $m_1 \neq m_2$. The F_0 parameters therefore represent the isotropic part of the coulombic repulsion between a pair of electrons. The $c^0(m_1, m_2)$ are relevant to all equivalent-electron configurations, as they will always satisfy the parity and triangle selection rules for equivalent electrons.

For non-zero values of k, the c^k are often fractions involving square roots. The 5 in front of c^2 in the table is the common denominator, d_k, to avoid fractions as entries in the table. Since the two-electron electrostatic interaction involves the product of two c^k integrals, it is convenient to re-define the Slater parameters so that they are divided by the square of this common denominator, shown in equation (6.45). In order to preserve the interaction energies, the product of the angular integrals therefore needs to be multiplied by the square of the common denominator, leaving them all conveniently as integers. The re-defined Slater parameters have their rank, k, written as a subscript instead of a superscript:

$$F_k = \frac{1}{d_k^2} F^k. \tag{6.45}$$

Exercise 95

The $M_S = 0, M_L = 2$ state of p^2 consists of a single microstate, $(1^+, 1^-)$, i.e. $m_l(1) = m_l(2) = 1$, $m_s(1) = +\frac{1}{2}$, $m_s(2) = -\frac{1}{2}$. Calculate the direct-coulombic and exchange-coulombic energies in terms of the F_k Slater parameters, and therefore the total electrostatic energy of interaction between the two electrons. Deduce the energy of the ^{1}D term of p^2.

Exercise 96

Calculate the total electrostatic energy of interaction between the two electrons in the ^{3}P term of p^2.

An M_S and M_L state of a two-electron term is often associated with more than one microstate of the uncoupled electrons. The energy of the $M_S M_L$ state is the sum of the energies of its component microstates. Since the energies of the $M_S M_L$ states are all diagonal terms of the Hamiltonian matrix, the rule about the summing of the component microstate energies is known as the diagonal sum rule [28].

We need to apply the diagonal sum rule to work out the electrostatic interaction energy of the ^{1}S term in p^2. Table 6.3 groups the microstates of p^2 according to their M_S and M_L values. We see that there are three microstates associated with the $M_S = 0, M_L = 0$ state. Each is associated

Table 6.3. The microstates of p^2 grouped according to their M_S and M_L values. The + and − superscripts denote the up and down projections of the electron spin.

	$M_S = 1$	$M_S = 0$	$M_S = -1$
$M_L = 2$		$(1^+, 1^-)$	
$M_L = 1$	$(1^+, 0^+)$	$(1^+, 0^-)(1^-, 0^+)$	$(1^-, 0^-)$
$M_L = 0$	$(1^+, -1^+)$	$(1^+, -1^-)(1^-, -1^+)(0^+, 0^-)$	$(1^-, -1^-)$
$M_L = -1$	$(-1^+, 0^+)$	$(-1^+, 0^-)(-1^-, 0^+)$	$(-1^-, 0^-)$
$M_L = -2$		$(-1^+, -1^-)$	

with one of the terms of p^2. The sum of their energies will be equal to the sum of the energies of the three terms.

Exercise 97

Calculate the total electrostatic energy of interaction between the two electrons in the ^1S term of p^2. Are the results consistent with Hund's rules (Section 4.6)? Show how this model predicts that the ^1S–^1D energy gap is 50% greater than the ^1D–^3P energy gap.

6.4.4 *Calculation of term energies for $p^{n>2}$*

We saw in Section 4.6 that the configurations p^n and p^{6-n} have the same Russell–Saunders terms, which should (assuming Hund's rules apply) be ordered in energy in the same way. We saw that the electrostatic energies of the terms of p^2 all contained a common F_0 parameter, and only differed in energy in their F_2 coefficients. In p^4 the terms have F_2 coefficients of the same magnitude and sign as they do in p^2, which accounts for the same relative ordering of the terms in energy. Even though we consider p^4 to be two positive electron 'holes', the sign of F_2 is the same as it is in p^2 because the electrostatic energy is a two-electron interaction, so the sign change from a 'positive electron' occurs twice, cancelling itself out.

In contrast, instead of a common F_0 in p^2, the terms in p^4 all have a common $6F_0$. This reflects the fact that there are six ways of choosing pairs of electrons from four p electrons. In general there are $n(n-1)/2$ ways of choosing pairs of electron from n electrons, and this factor can always be found in front of the F_0 parameter in the energies of Russell–Saunders terms from configurations of n equivalent electrons.

Once the values of the F_0 and F_2 parameters have been worked out from the observed term energies, they will be expected to be different for different elements. This reflects how the electrostatic interactions depend on the radial characteristics of all the orbitals occupied with electrons, which will be different for different elements (with different nuclear charges and different numbers of electrons).

The calculation of the angular coefficients of the Slater parameters for the p^3 configuration is more involved. In the application of the scheme for pairs of electrons to uncoupled three-electron microstates ($e_1e_2e_3$), the three possible electron pairs (e_1e_2, e_1e_3 and e_2e_3) need to be summed over for both direct-coulombic and exchange-coulombic energies.

Exercise 98

Find the difference in energy between the ground (^{4}S) and first-excited (^{2}D) terms in the p^3 configuration.

Exercise 99

A value for the electrostatic Slater parameter F_2 can be established from the energy gap between the ground and first-excited terms in carbon (p^2), nitrogen (p^3) and oxygen (p^4). The energy gaps, in wavenumbers, between the ground and first-excited term barycentres are: 10164 (C), 19226 (N) and 15790 (O) [31]. Find these F_2 values and suggest a trend in F_2 across the period. Why can analogous F_2 values not be determined for boron and fluorine atoms?

When there are greater than two electrons in subshells of higher angular momentum, the calculation is not only more tedious using our microstates diagonal sum method but will also produce more than one term with the same S and L quantum numbers. More powerful methods are needed for these configurations that are beyond the scope of this book. Interested readers should consult Condon and Shortley [29], Griffith [28], Wybourne [72] and Judd [96].

6.4.5 *Calculation of spin–orbit coupling energies*

For an individual electron, the energy operator for the spin–orbit interaction is $\xi(r)\mathbf{s}.\mathbf{l}$, where $\xi(r)$ is a function of only radial coordinates. Since the scalar product of two angular momenta will have units of $\hbar^2$, the quantity $\xi(r)\hbar^2$ should have units of energy. This radial contribution to the energy of the spin–orbit interaction is collected in the constant ζ_{nl}, which depends on n and l as these quantum numbers both influence the radial wavefunction. The expression for ζ_{nl}, which is shown below, is derived in [10]:

$$\zeta_{nl} = \frac{\alpha^2 R_\infty Z^4}{n^3 l(l+\frac{1}{2})(l+1)}. \tag{6.46}$$

Every term on the right-hand side of equation (6.46) is dimensionless apart from R_∞, the Rydberg constant with the mass (rather than the reduced mass) of the electron, which was described in Section 2.8. The Z^4 term is the reason that the spin–orbit interaction is often said to increase with the fourth power of atomic number, though this is only strictly true for hydrogenic atoms. α is the fine-structure constant and n and l are the quantum numbers of the electron.

ζ_{nl} is taken to be the adjustable radial parameter that accompanies the angular calculation of the spin–orbit energies, just as the Slater parameters were the adjustable radial parameters for the angular electrostatic energy calculations. Given that $\hat{\mathbf{j}}\psi = \hbar m_j\psi$, where $\mathbf{j}$ is any angular momentum vector, we can find the spin–orbit interaction energy, $E_{\text{s}-\text{o}}$, for a single electron as follows, assuming the wavefunction is normalised:

$$\begin{aligned} E_{\text{s}-\text{o}} &= \langle \psi_{m_s m_l} \mid \xi(r)\mathbf{s}.\mathbf{l} \mid \psi_{m_s m_l} \rangle \\ &= \xi(r)\hbar^2 m_s m_l \langle \psi_{m_s m_l} | \psi_{m_s m_l} \rangle \\ &= \zeta_{nl} m_s m_l. \end{aligned} \tag{6.47}$$

This process may be summed over all the electrons in an uncoupled microstate. The m_s and m_l do not affect one another in this calculation as the two forms of angular momentum commute.

In Section 4.5 the total angular momentum vector, $\mathbf{J}$, was defined as the vector sum $\mathbf{S}+\mathbf{L}$. A multi-electron parameter, λ_{SL}, may also be defined for a Russell–Saunders term, ^{2S+1}L, by expanding $\mathbf{J}^2$:

$$\mathbf{J}^2 = (\mathbf{S} + \mathbf{L})^2 = \mathbf{S}^2 + \mathbf{L}^2 + 2\mathbf{S}.\mathbf{L}$$

$$\mathbf{S}.\mathbf{L} = \frac{1}{2}(\mathbf{J}^2 - \mathbf{S}^2 - \mathbf{L}^2). \tag{6.48}$$

Since $\hat{\mathbf{J}^2}\psi = \hbar^2 J(J+1)\psi$, where $\mathbf{J}$ is any angular momentum vector, we can find the spin–orbit interaction energy, $E_{\text{s}-\text{o}}$, for a Russell–Saunders term as follows, assuming the wavefunction is normalised:

$$\begin{aligned} E_{\text{s}-\text{o}} &= \langle \psi_{SL} \mid \xi(r)\mathbf{S}.\mathbf{L} \mid \psi_{SL} \rangle \\ &= \xi(r)\frac{\hbar^2}{2}(J(J+1) - S(S+1) - L(L+1))\langle \psi_{SL} | \psi_{SL} \rangle \\ &= \lambda_{SL}\frac{1}{2}(J(J+1) - S(S+1) - L(L+1)). \end{aligned} \tag{6.49}$$

We can see that the result for the spin–orbit interaction energy will always be zero for singlet states: when $S = 0$, it follows that $J = L$. This will cause $E_{\text{s}-\text{o}}$ to vanish.

Exercise 100

Show that the Landé interval rule for the difference in energy between J and $J-1$ levels within a particular Russell–Saunders term, ^{2S+1}L, is $\lambda_{SL}J$.

It is advantageous to express all the spin–orbit interaction energies in a configuration in terms of a single parameter, ζ_{nl}, rather than separate λ_{SL} parameters for every term. The smaller the number of adjustable parameters, the more meaningful will be their values when fitted to experimental data. This is achieved in a similar way to the calculation of the coefficients of the Slater parameters, using uncoupled microstates as described above.

If only a single microstate corresponds to an $M_S M_L$ combination, then the sum of the spin–orbit energies for the individual electrons gives the spin–orbit energy for the J level of the term, where the J is equal to $M_J = M_S + M_L$. If more than one microstate corresponds to an $M_S M_L$ combination required to represent a term then the diagonal sum rule is employed: the sum of the spin–orbit energies of all the electrons in the uncoupled microstates is equal to the sum of the spin–orbit energies for the J level of the terms described by the microstates.

Exercise 101

Find the λ spin–orbit parameter for the ground-state term of gaseous iron atoms in terms of ζ_{3d}.

Exercise 102

Deduce the energies of all the J levels in the ^{5}D multiplet of d^6 in terms of $\lambda_{^5\mathrm{D}}$. Show that the average spin–orbit energy of these levels is zero.

Exercise 103

Use the four following observed energy levels above the ^{5}D$_4$ ground state of Fe atoms and the Landé interval rule to deduce a value of ζ_{3d} in wavenumbers for each interval. Comment on the result. The energies, in wavenumbers, are as follows: 415.9 (^{5}D$_3$), 704.0 (^{5}D$_2$), 888.1 (^{5}D$_1$) and 978.1 (^{5}D$_0$) [41].

In order for a good agreement with the Landé interval rule to be observed, the spin–orbit interaction needs to be large enough so that other magnetic interactions (see Section 4.8) have little impact, but not so large that second-order effects are seen between adjacent multiplets. These interfering magnetic interactions account for the occasional reversal of the expected J levels observed in multiplets.

Exercise 104

For the ^{3}F and ^{3}P terms in d^2, find the λ spin–orbit parameters in terms of the ζ_{3d} parameter.

Exercise 105

Use the following observed energy levels to determine four values for the ζ_{3d} parameter in titanium (3d^2 4s^2). The energies, in wavenumbers, are: 0.0 (^{3}F$_2$), 170.1 (^{3}F$_3$), 386.9 (^{3}F$_4$), 8436.6 (^{3}P$_0$), 8492.4 (^{3}P$_1$) and 8602.4 (^{3}P$_2$) [31]. Compare these ζ_{3d} parameters to those found earlier for iron. Does the increase across the period resemble a Z^4 relationship?

The proportionality of ζ values to Z^4 is most apparent in hydrogenic atoms — the context in which the relation was derived. The spin–orbit splitting of the ^{2}P term of 2p^1 is recorded in Moore's tables for hydrogenic atoms up to O^{7+}. Application of equations (6.47) and (6.49) reveals that $\lambda_{^2\mathrm{P}} = \zeta_{2p}$. According to the Landé interval rule, the energy gap between $^2\mathrm{P}_{3/2}$ and $^2\mathrm{P}_{1/2}$ is $\frac{3}{2}\lambda_{^2\mathrm{P}} = \frac{3}{2}\zeta_{2p}$. The ζ_{2p} (zeta) values, measured in wavenumbers, are plotted against Z^4 in Figure 6.9. A linear relationship is clearly observed.

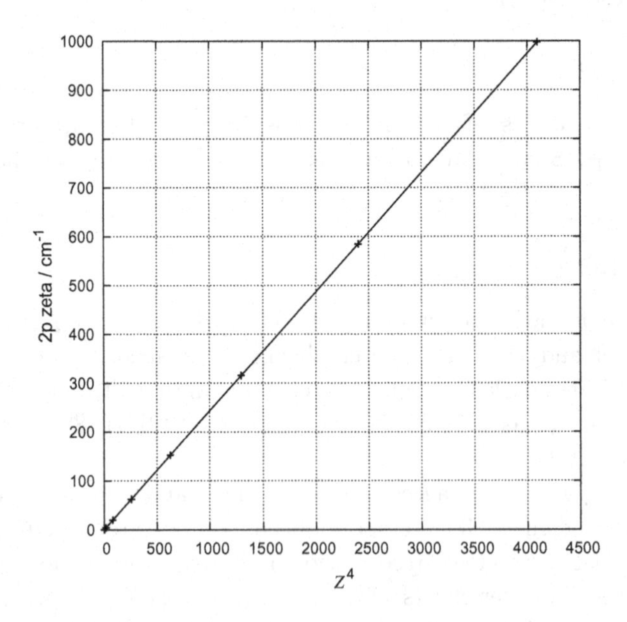

Figure 6.9. Observed hydrogenic atom ζ_{2p} (zeta) values from hydrogen to oxygen plotted against Z^4 (data from [31]).

Exercise 106

Find the spin–orbit parameter for every term in the configuration p^3. Comment on the result.

6.4.6 *d- and f-block elements*

The atomic spectra of d- and f-block elements are complicated by multiple electron configurations overlapping in energy. In d-block elements, there are same-parity configurations overlapping with the ground configuration. $d^n s^2$, $d^{n+1}s$ and d^{n+2} configurations are all even-parity. The Hamiltonian operator is a scalar because energy is a scalar, and therefore is even-parity. This means it can operate between same-parity configurations in a process known as configuration interaction. These interactions are most significant when the interacting configurations are close in energy. The relative average-energies observed for these even-parity configurations in the three rows of the d block are given in Figures 3.11 to 3.13.

Since the electric dipole operator is odd-parity, the important transitions are odd-parity, of the type s→p and d→p (and f→d in the f block), for example. Excited states in configurations of the same parity as the ground configuration can therefore be metastable. The spectra of ionised transition metal atoms are often simpler than for neutral atoms as the d subshell is further below the valence s subshell in energy when the atom is ionised. Atoms in groups 11 and 12 with analogous electron configurations to s-block elements have some similar features to them in their spectra too.

Spin–orbit coupling, jj coupling and other magnetic effects are very significant in the third row of the d block and the second row of the f block. The 14 lanthanides before the third row of the d block make for a particularly large increase in proton number in the third row of the d block. The additional nuclear attraction significantly increases these magnetic effects.

With ns, $(n-1)$d and $(n-2)$f subshells all close in energy in the f block, these atoms have multiple overlapping configurations, making spectral assignment difficult for the neutral atoms. Assignment is simpler for the 3+ and higher charged ions, as the f subshell is lower in energy relative to the outer s and d subshells in this ionised form. With their high orbital angular momentum and lack of any radial nodes, the 4f orbitals are quite core-like. Even in aqueous solution and in insulating crystals, tripositive lanthanide ions generate quite sharp spectra. Spin–orbit splitting of terms is clearly resolved. In insulating crystals, the crystal-field splitting of lines is small compared to electrostatic interactions and spin–orbit coupling. The 5f orbitals have a radial node and so are less core-like in actinides compared to the 4f orbitals in lanthanides. The actinide spectra therefore resemble those of the third-row d block more than the lanthanides.

With the high angular momentum of f electrons and their sharp lines even in condensed phases, very many levels are observed in lanthanide spectra. Lanthanides are therefore good candidates for fitting radial parameters to experimental data and comparing different schemes of parametrisation [97, 98].

Appendix A

Worked Solutions to Exercises

Chapter 1: Fundamentals

Exercise 1 Starting at the right-hand side, the (2×2) matrix operates on the (2×1) complex vector to produce another (2×1) complex vector:

$$\begin{bmatrix} m & x+\mathrm{i}y \\ x-\mathrm{i}y & n \end{bmatrix} \begin{bmatrix} a+\mathrm{i}b \\ c+\mathrm{i}d \end{bmatrix} = \begin{bmatrix} m(a+\mathrm{i}b)+(x+\mathrm{i}y)(c+\mathrm{i}d) \\ (x-\mathrm{i}y)(a+\mathrm{i}b)+n(c+\mathrm{i}d) \end{bmatrix}.$$

This (2×1) complex vector is pre-multiplied by the first (1×2) complex vector to give a (1×1) scalar.

$$\begin{aligned} &\begin{bmatrix} a-\mathrm{i}b & c-\mathrm{i}d \end{bmatrix} \begin{bmatrix} m(a+\mathrm{i}b)+(x+\mathrm{i}y)(c+\mathrm{i}d) \\ (x-\mathrm{i}y)(a+\mathrm{i}b)+n(c+\mathrm{i}d) \end{bmatrix} \\ &= m(a+\mathrm{i}b)(a-\mathrm{i}b)+(a-\mathrm{i}b)(x+\mathrm{i}y)(c+\mathrm{i}d) \\ &\quad +(c-\mathrm{i}d)(x-\mathrm{i}y)(a+\mathrm{i}b)+n(c+\mathrm{i}d)(c-\mathrm{i}d) \end{aligned}$$

In order for the (2×2) matrix to be Hermitian, the (1×1) scalar must be real. The first and last terms are evidently real. The second and third terms, when multiplied out, will each contain three terms in imaginary i and one in i^3. The two factors of i^3 are $-bdy$ and bdy and so cancel out. The six factors of i are

$$-bcx + acy + adx - adx - acy + bcx.$$

Since a, b, c, d, x, and y are real, the (2×2) matrix must be Hermitian.

Exercise 2 Operating first with the right-hand operator gives the following expression:

$$
\begin{aligned}
[\hat{x}, \hat{p_x}]\psi &= \left(x \times \frac{h}{2\pi\mathrm{i}}\frac{\partial\psi}{\partial x}\right) - \left(\frac{h}{2\pi\mathrm{i}}\frac{\partial(x \times \psi)}{\partial x}\right) \\
&= \frac{h}{2\pi\mathrm{i}}\left(x\frac{\partial\psi}{\partial x} - x\frac{\partial\psi}{\partial x} - \psi\right) = -\frac{h}{2\pi\mathrm{i}}\psi = \frac{\mathrm{i}h}{2\pi}\psi = \mathrm{i}\hbar\psi.
\end{aligned}
$$

Hence the commutator is equal to $\frac{\mathrm{i}h}{2\pi}$ or $\mathrm{i}\hbar$. A finite commutator demonstrates that the position and momentum of an electron may not simultaneously be known exactly.

Exercise 3 The two operators commute if $\hat{A}\hat{B}\psi = \hat{B}\hat{A}\psi$. We operate from the right and bear in mind that scalars commute with operators:

$$
\hat{A}\hat{B}\psi = \hat{A}b\psi = b\hat{A}\psi = \hat{B}\hat{A}\psi.
$$

Exercise 4 Substituting in the expressions for $|A\rangle$ and $|B\rangle$ and remembering that $\langle A| = |A\rangle^*$ gives the following:

$$
\begin{aligned}
(a - \mathrm{i}b)(x + \mathrm{i}y) &= ax + \mathrm{i}ay - \mathrm{i}bx + by \\
\{(x - \mathrm{i}y)(a + \mathrm{i}b)\}^* &= \{ax + \mathrm{i}bx - \mathrm{i}ay + by\}^* \\
&= ax - \mathrm{i}bx + \mathrm{i}ay + by.
\end{aligned}
$$

This demonstrates that $\langle A|B\rangle = \langle B|A\rangle^*$.

Chapter 2: Orbitals in the Hydrogen Atom

Exercise 5 Remembering the convention for anti-clockwise rotation in a right-handed coordinate system, the approach taken to $\hat{l_z}$ gives the following Cartesian expressions for $\hat{l_x}$ and $\hat{l_y}$, which display a cyclic permutation of axis labels:

$$
\begin{aligned}
\hat{l_x} &= \frac{\hbar}{\mathrm{i}}\left(y\frac{\partial}{\partial z} - z\frac{\partial}{\partial y}\right) \\
\hat{l_y} &= \frac{\hbar}{\mathrm{i}}\left(z\frac{\partial}{\partial x} - x\frac{\partial}{\partial z}\right).
\end{aligned}
\tag{A.1}
$$

Exercise 6 Substituting the expanded forms of the $\hat{l}_x$ and $\hat{l}_y$ operators from equations (A.1) into the commutator leads to the following expansion:

$$\begin{aligned}
\hat{l}_x\hat{l}_y &= -\hbar^2\left(y\frac{\partial}{\partial z} - z\frac{\partial}{\partial y}\right)\left(z\frac{\partial}{\partial x} - x\frac{\partial}{\partial z}\right)\\
&= -\hbar^2\left(y\frac{\partial}{\partial z}z\frac{\partial}{\partial x} - y\frac{\partial}{\partial z}x\frac{\partial}{\partial z} - z\frac{\partial}{\partial y}z\frac{\partial}{\partial x} + z\frac{\partial}{\partial y}x\frac{\partial}{\partial z}\right)\\
\hat{l}_y\hat{l}_x &= -\hbar^2\left(z\frac{\partial}{\partial x} - x\frac{\partial}{\partial z}\right)\left(y\frac{\partial}{\partial z} - z\frac{\partial}{\partial y}\right)\\
&= -\hbar^2\left(z\frac{\partial}{\partial x}y\frac{\partial}{\partial z} - z\frac{\partial}{\partial x}z\frac{\partial}{\partial y} - x\frac{\partial}{\partial z}y\frac{\partial}{\partial z} + x\frac{\partial}{\partial z}z\frac{\partial}{\partial y}\right).
\end{aligned}$$

The second and third terms of each expansion cancel out in a simple way when $\hat{l}_y\hat{l}_x$ is subtracted from $\hat{l}_x\hat{l}_y$ since the differentials do not involve the variables that multiply them. This leaves the following expression for the commutator:

$$\begin{aligned}
[\hat{l}_x,\, \hat{l}_y] &= \hat{l}_x\hat{l}_y - \hat{l}_y\hat{l}_x\\
&= -\hbar^2\left(y\frac{\partial}{\partial z}z\frac{\partial}{\partial x} + z\frac{\partial}{\partial y}x\frac{\partial}{\partial z} - z\frac{\partial}{\partial x}y\frac{\partial}{\partial z} - x\frac{\partial}{\partial z}z\frac{\partial}{\partial y}\right)\\
&= -\hbar^2\left(y\frac{\partial}{\partial x} + yz\frac{\partial^2}{\partial z\partial x} + zx\frac{\partial^2}{\partial y\partial z} - zy\frac{\partial^2}{\partial x\partial z} - x\frac{\partial}{\partial y} - xz\frac{\partial^2}{\partial z\partial y}\right)\\
&= -\hbar^2\left(y\frac{\partial}{\partial x} - x\frac{\partial}{\partial y}\right) = i\hbar\hat{l}_z.
\end{aligned}$$

The product rule of differentiation applies to the first and last terms in the first bracket, leading to further cancellation.

Exercise 7 The other commutators are given by cyclic permutation symmetry, and the fact that exchanging the operators in the commutator introduces a minus sign. The results are given below:

$$\begin{aligned}
[\hat{l}_x,\, \hat{l}_y] &= \mathrm{i}\hbar\hat{l}_z \quad [\hat{l}_y,\, \hat{l}_x] = -\mathrm{i}\hbar\hat{l}_z\\
[\hat{l}_y,\, \hat{l}_z] &= \mathrm{i}\hbar\hat{l}_x \quad [\hat{l}_z,\, \hat{l}_y] = -\mathrm{i}\hbar\hat{l}_x\\
[\hat{l}_z,\, \hat{l}_x] &= \mathrm{i}\hbar\hat{l}_y \quad [\hat{l}_x,\, \hat{l}_z] = -\mathrm{i}\hbar\hat{l}_y.
\end{aligned} \tag{A.2}$$

Exercise 8 First, equations (2.2) need to be separately differentiated with respect to x, y and z. The results are then expressed in terms of the spherical polar variables to find the differentials of spherical polar variables with respect to Cartesian variables that are required in equation (2.15). This yields the following results:

$$\frac{\partial r}{\partial x} = \sin\theta\cos\varphi \quad \frac{\partial \theta}{\partial x} = \frac{\cos\theta\cos\varphi}{r} \quad \frac{\partial \varphi}{\partial x} = \frac{-\sin\varphi}{r\sin\theta}$$

$$\frac{\partial r}{\partial y} = \sin\theta\sin\varphi \quad \frac{\partial \theta}{\partial y} = \frac{\cos\theta\sin\varphi}{r} \quad \frac{\partial \varphi}{\partial y} = \frac{\cos\varphi}{r\sin\theta}$$

$$\frac{\partial r}{\partial z} = \cos\theta \quad \frac{\partial \theta}{\partial z} = \frac{-\sin\theta}{r} \quad \frac{\partial \varphi}{\partial z} = 0. \tag{A.3}$$

These expressions and equations (2.2) are then substituted into equation (2.15) and equations (2.4) and (A.1) to give the following expressions for the operators $\hat{l_x}$, $\hat{l_y}$ and $\hat{l_z}$ in spherical polar coordinates:

$$\hat{l_x} = \frac{\hbar}{\mathrm{i}}\left(-\sin\varphi\frac{\partial}{\partial\theta} - \cot\theta\cos\varphi\frac{\partial}{\partial\varphi}\right) \tag{A.4a}$$

$$\hat{l_y} = \frac{\hbar}{\mathrm{i}}\left(\cos\varphi\frac{\partial}{\partial\theta} - \cot\theta\sin\varphi\frac{\partial}{\partial\varphi}\right) \tag{A.4b}$$

$$\hat{l_z} = \frac{\hbar}{\mathrm{i}}\frac{\partial}{\partial\varphi}. \tag{A.4c}$$

Exercise 9 We use the expression for $\hat{l_z}$ in spherical polar coordinates and operate on $\mathrm{e}^{\mathrm{i}m\varphi}$, which confirms that equation (2.6) is true:

$$\hat{l_z}\psi = \frac{\hbar}{\mathrm{i}}\frac{\partial\left(\mathrm{e}^{\mathrm{i}m\varphi}\right)}{\partial\varphi} = \frac{\hbar}{\mathrm{i}}\mathrm{i}m\mathrm{e}^{\mathrm{i}m\varphi} = \hbar m\psi.$$

Exercise 10 Each principal shell contains n subshells with l taking values from 0 to $n-1$. Each of these subshells contains $2l+1$ orbitals. This leads to the sequence of orbitals per shell as 1, $1+3$, $1+3+5$, $\ldots$, which can be simplified to n^2.

Exercise 11 If $\langle -m|m\rangle = 0$ then $|-m\rangle$ and $|m\rangle$ are orthogonal.

$$\langle -m|m\rangle = \left\langle \frac{e^{-im\varphi}}{\sqrt{2\pi}} \middle| \frac{e^{im\varphi}}{\sqrt{2\pi}} \right\rangle = \frac{1}{2\pi}\int_0^{2\pi} e^{im\varphi}e^{im\varphi}\, d\varphi$$

$$= \frac{1}{2\pi}\int_0^{2\pi} e^{i2m\varphi}\, d\varphi = \frac{1}{i4\pi m}\left[e^{i2m\varphi}\right]_0^{2\pi} = 0$$

Exercise 12 We need to show that $\langle P_1^1|P_1^0\rangle = 0$. To integrate these functions over all space we need to use the Jacobian $\sin\theta$:

$$\langle P_1^1|P_1^0\rangle = \int_0^{\pi} \sin\theta\cos\theta\,\sin\theta\, d\theta = \left[\frac{1}{3}\sin^3\theta\right]_0^{\pi} = 0.$$

Exercise 13 We need to square P_2^0 and integrate between 0 and π with the Jacobian $\sin\theta$ (as θ is part of the spherical polar coordinate system):

$$\int_0^{\pi} N_\theta^2\left(P_2^0\right)^2 \sin\theta\, d\theta = N_\theta^2 \int_0^{\pi} (9\cos^4\theta - 6\cos^2\theta + 1)\sin\theta\, d\theta$$

$$= N_\theta^2\left[-\frac{9}{5}\cos^5\theta + 2\cos^3\theta - \cos\theta\right]_0^{\pi} = N_\theta^2 . \frac{8}{5}$$

$$= 1$$

$$N_\theta = \sqrt{\frac{5}{8}} = \frac{\sqrt{10}}{4}.$$

Exercise 14 We need to show that

$$\langle e^{im\varphi} + e^{-im\varphi}|e^{im\varphi} - e^{-im\varphi}\rangle = 0.$$

Taking the complex conjugate of the bra function and multiplying out the brackets gives the following:

$$(e^{-im\varphi} + e^{im\varphi})(e^{im\varphi} - e^{-im\varphi}) = 1 - e^{-i2m\varphi} + e^{i2m\varphi} - 1.$$

Integrating this function with respect to φ between 0 and 2π gives 0, shown below:

$$\frac{1}{i2m}\left[e^{-i2m\varphi} + e^{i2m\varphi}\right]_0^{2\pi} = 0.$$

Exercise 15 The derivatives of spherical polar variables with respect to Cartesian variables from equations (A.3) need to be converted back into Cartesian variables, differentiated again and converted back into spherical polar variables using equations (2.2). This gives the following results:

$$\frac{\partial^2 r}{\partial x^2} = \frac{\sin^2\theta\sin^2\varphi + \cos^2\theta}{r} \tag{A.5}$$

$$\frac{\partial^2 \theta}{\partial x^2} = \frac{\cos\theta}{r^2\sin\theta}\left(\sin^2\varphi - 2\sin^2\theta\cos^2\varphi\right) \tag{A.6}$$

$$\frac{\partial^2 \varphi}{\partial x^2} = \frac{2\sin\varphi\cos\varphi}{r^2\sin^2\theta}. \tag{A.7}$$

These equations are substituted into equation (2.16), giving the following result:

$$\begin{aligned}\frac{\partial^2\psi}{\partial x^2} &= \frac{\sin^2\theta\sin^2\varphi + \cos^2\theta}{r}\frac{\partial\psi}{\partial r} + \sin^2\theta\cos^2\varphi\frac{\partial^2\psi}{\partial r^2}\\ &\quad + \frac{\cos\theta}{r^2\sin\theta}\left(\sin^2\varphi - 2\sin^2\theta\cos^2\varphi\right)\frac{\partial\psi}{\partial\theta} + \frac{\cos^2\theta\cos^2\varphi}{r^2}\frac{\partial^2\psi}{\partial\theta^2}\\ &\quad + \frac{2\sin\varphi\cos\varphi}{r^2\sin^2\theta}\frac{\partial\psi}{\partial\varphi} + \frac{\sin^2\varphi}{r^2\sin^2\theta}\frac{\partial^2\psi}{\partial\varphi^2}.\end{aligned} \tag{A.8}$$

Exercise 16 The derivatives of spherical polar variables with respect to Cartesian variables from equations (A.3) need to be converted back into Cartesian variables, differentiated again and converted back into spherical polar variables using equations (2.2). This gives the following results:

$$\frac{\partial^2 r}{\partial y^2} = \frac{\sin^2\theta\cos^2\varphi + \cos^2\theta}{r} \quad \frac{\partial^2 r}{\partial z^2} = \frac{\sin^2\theta}{r} \tag{A.9}$$

$$\frac{\partial^2 \theta}{\partial y^2} = \frac{\cos\theta}{r^2\sin\theta}\left(\cos^2\varphi - 2\sin^2\theta\sin^2\varphi\right) \quad \frac{\partial^2\theta}{\partial z^2} = \frac{2\sin\theta\cos\theta}{r^2} \tag{A.10}$$

$$\frac{\partial^2 \varphi}{\partial y^2} = \frac{-2\sin\varphi\cos\varphi}{r^2\sin^2\theta} \quad \frac{\partial^2\varphi}{\partial z^2} = 0. \tag{A.11}$$

These equations are substituted into equation (2.16), giving the following result:

$$\frac{\partial^2\psi}{\partial y^2} = \frac{\sin^2\theta\cos^2\varphi + \cos^2\theta}{r}\frac{\partial\psi}{\partial r} + \sin^2\theta\sin^2\varphi\frac{\partial^2\psi}{\partial r^2}$$

$$+\frac{\cos\theta}{r^2\sin\theta}\left(\cos^2\varphi - 2\sin^2\theta\sin^2\varphi\right)\frac{\partial\psi}{\partial\theta} + \frac{\cos^2\theta\sin^2\varphi}{r^2}\frac{\partial^2\psi}{\partial\theta^2}$$

$$-\frac{2\sin\varphi\cos\varphi}{r^2\sin^2\theta}\frac{\partial\psi}{\partial\varphi} + \frac{\cos^2\varphi}{r^2\sin^2\theta}\frac{\partial^2\psi}{\partial\varphi^2} \tag{A.12a}$$

$$\frac{\partial^2\psi}{\partial z^2} = \frac{\sin^2\theta}{r}\frac{\partial\psi}{\partial r} + \cos^2\theta\frac{\partial^2\psi}{\partial r^2} + \frac{2\sin\theta\cos\theta}{r^2}\frac{\partial\psi}{\partial\theta} + \frac{\sin^2\theta}{r^2}\frac{\partial^2\psi}{\partial\theta^2}. \tag{A.12b}$$

Collecting terms in the derivatives with respect to spherical polar coordinates for equations (A.8) and (A.12) yields the following result:

$$\nabla^2\psi = \frac{\partial^2\psi}{\partial r^2} + \frac{2}{r}\frac{\partial\psi}{\partial r} + \frac{1}{r^2}\frac{\partial^2\psi}{\partial\theta^2} + \frac{\cos\theta}{r^2\sin\theta}\frac{\partial\psi}{\partial\theta} + \frac{1}{r^2\sin^2\theta}\frac{\partial^2\psi}{\partial\varphi^2}. \tag{A.13}$$

Expansion of equation (2.18) gives the same result.

Exercise 17 We need to show that the determinant of the Jacobian matrix is $r^2\sin\theta$. The differential elements of the matrix are found by differentiating equations (2.2):

$$\begin{aligned}
\frac{\partial x}{\partial r} &= \sin\theta\cos\varphi & \frac{\partial y}{\partial r} &= \sin\theta\sin\varphi & \frac{\partial z}{\partial r} &= \cos\theta\\
\frac{\partial x}{\partial\theta} &= r\cos\theta\cos\varphi & \frac{\partial y}{\partial\theta} &= r\cos\theta\sin\varphi & \frac{\partial z}{\partial\theta} &= -r\sin\theta\\
\frac{\partial x}{\partial\varphi} &= -r\sin\theta\sin\varphi & \frac{\partial y}{\partial\varphi} &= r\sin\theta\cos\varphi & \frac{\partial z}{\partial\varphi} &= 0.
\end{aligned}$$

It is quickest to expand the determinant in the third row as one of the elements is zero. This leads to the following expression:

$$\frac{\partial z}{\partial r}\left(\frac{\partial x}{\partial\theta}\frac{\partial y}{\partial\varphi} - \frac{\partial x}{\partial\varphi}\frac{\partial y}{\partial\theta}\right) - \frac{\partial z}{\partial\theta}\left(\frac{\partial x}{\partial r}\frac{\partial y}{\partial\varphi} - \frac{\partial x}{\partial\varphi}\frac{\partial y}{\partial r}\right).$$

Substitution of the differentials into this expression gives $r^2\sin\theta$ as desired.

Exercise 18 The operators given in equations (A.4c) are multiplied out taking special care over the order of the terms (as operators are not, in general, commutative). This yields the following results:

$$\hat{l}_x^2 = -\hbar^2 \left\{ \sin^2\varphi \frac{\partial^2}{\partial\theta^2} + \cot^2\theta\cos^2\varphi \frac{\partial^2}{\partial\varphi^2} + 2\cot\theta\sin\varphi\cos\varphi \frac{\partial^2}{\partial\theta\partial\varphi} \right.$$
$$\left. - \sin\varphi\cos\varphi\left(\cot^2\theta + \frac{1}{\sin^2\theta}\right)\frac{\partial}{\partial\varphi} + \cot\theta\cos^2\theta\frac{\partial}{\partial\theta} \right\}$$

$$\hat{l}_y^2 = -\hbar^2 \left\{ \cos^2\varphi \frac{\partial^2}{\partial\theta^2} + \cot^2\theta\sin^2\varphi \frac{\partial^2}{\partial\varphi^2} - 2\cot\theta\sin\varphi\cos\varphi \frac{\partial^2}{\partial\theta\partial\varphi} \right.$$
$$\left. + \sin\varphi\cos\varphi\left(\cot^2\theta + \frac{1}{\sin^2\theta}\right)\frac{\partial}{\partial\varphi} + \cot\theta\sin^2\theta\frac{\partial}{\partial\theta} \right\}$$

$$\hat{l}_z^2 = -\hbar^2 \frac{\partial^2}{\partial\varphi^2}.$$

Adding these equations and collecting the terms in the differentials with respect to spherical polar variables gives the following result:

$$\hat{l}^2 = -\hbar^2 \left(\frac{\partial^2}{\partial\theta^2} + \frac{\cos\theta}{\sin\theta}\frac{\partial}{\partial\theta} + \frac{1}{\sin^2\theta}\frac{\partial^2}{\partial\varphi^2} \right).$$

We can rearrange this expression using the following relation:

$$\frac{1}{\sin\theta}\frac{\partial}{\partial\theta}\sin\theta\frac{\partial}{\partial\theta} = \frac{1}{\sin\theta}\left(\sin\theta\frac{\partial^2}{\partial\theta^2} + \cos\theta\frac{\partial}{\partial\theta}\right)$$
$$= \frac{\partial^2}{\partial\theta^2} + \frac{\cos\theta}{\sin\theta}\frac{\partial}{\partial\theta}.$$

We therefore get this more recognisable result:

$$\hat{l}^2 = -\hbar^2 \left(\frac{1}{\sin\theta}\frac{\partial}{\partial\theta}\sin\theta\frac{\partial}{\partial\theta} + \frac{1}{\sin^2\theta}\frac{\partial^2}{\partial\varphi^2} \right). \tag{A.14}$$

Exercise 19 We investigate the effect of the operator acting on the ket wavefunction in case it returns just a constant multiplied by the wavefunction:

$$\begin{aligned}
\hat{l^2}(\sin^2\theta\cos 2\varphi) &= -\hbar^2\left(\frac{1}{\sin\theta}\frac{\partial}{\partial\theta}\sin\theta\frac{\partial}{\partial\theta}+\frac{1}{\sin^2\theta}\frac{\partial^2}{\partial\varphi^2}\right)(\sin^2\theta\cos 2\varphi)\\
&= -\hbar^2\left(\frac{1}{\sin\theta}\frac{\partial}{\partial\theta}(2\sin^2\theta\cos\theta\cos 2\varphi)-4\cos 2\varphi\right)\\
&= -\hbar^2(4\cos^2\theta\cos 2\varphi-2\sin^2\theta\cos 2\varphi-4\cos 2\varphi)\\
&= -\hbar^2\{4(1-\sin^2\theta)\cos 2\varphi-2\sin^2\theta\cos 2\varphi-4\cos 2\varphi\}\\
&= -\hbar^2(-6\sin^2\theta\cos 2\varphi).
\end{aligned}$$

We can see that the operator has returned $6\hbar^2$ multiplied by the angular function, which is in accordance with $\hbar^2 l(l+1)$.

Exercise 20 We just consider the radial coordinate and apply the Jacobian determinant to obtain the integral over all r that must give 1. The radial wavefunction is real so may simply be squared. For the purposes of this question, the normalisation constant is taken to be everything that is multiplied by the exponential function.

$$\int_0^\infty \left(N\mathrm{e}^{-\rho r/2}\right)^2 r^2\,\mathrm{d}r = 1$$

$$N^2\int_0^\infty \mathrm{e}^{-\rho r}\,r^2\,\mathrm{d}r = 1$$

Integration by parts is required. We need to use the following result:

$$\begin{aligned}
\int x^2\mathrm{e}^{-kx}\,\mathrm{d}x &= x^2\cdot -\frac{1}{k}\mathrm{e}^{-kx} - \int -\frac{1}{k}\mathrm{e}^{-kx}\cdot 2x\,\mathrm{d}x + c\\
&= x^2\cdot -\frac{1}{k}\mathrm{e}^{-kx} - \left(2x\cdot\frac{1}{k^2}\mathrm{e}^{-kx} - \int\frac{1}{k^2}\mathrm{e}^{-kx}\cdot 2\,\mathrm{d}x\right) + c\\
&= x^2\cdot -\frac{1}{k}\mathrm{e}^{-kx} - \left(2x\cdot\frac{1}{k^2}\mathrm{e}^{-kx} - \left(-\frac{2}{k^3}\mathrm{e}^{-kx}\right)\right) + c\\
&= -\mathrm{e}^{-kx}\left(\frac{x^2}{k}+\frac{2x}{k^2}+\frac{2}{k^3}\right)+c. \qquad \text{(A.15)}
\end{aligned}$$

We can now solve for N:

$$-N^2\left[\mathrm{e}^{-\rho r}\left(\frac{r^2}{\rho}+\frac{2r}{\rho^2}+\frac{2}{\rho^3}\right)\right]_0^\infty = 1$$

$$-N^2\left(0-\frac{2}{\rho^3}\right) = 1$$

$$\frac{2N^2}{\rho^3} = 1.$$

We now substitute in the expression for ρ (equation (2.27)), set n to 1 and solve for N:

$$N^2 = \frac{4Z^3}{a_0^3}$$

$$N = 2\left(\frac{Z}{a_0}\right)^{\frac{3}{2}}.$$

This has given us the desired result.

Exercise 21 Since there is no angular part of the 1s orbital wavefunction we can neglect the angular operators in equation (2.22) and just substitute in the radial wavefunction. The required expressions can be written in Dirac notation (equation (1.22)). The integral may be factorised into kinetic and potential energy parts:

$$\left\langle \psi_{1s} \left| \frac{-\hbar^2}{2\mu r^2}\left(\frac{\partial}{\partial r}r^2\frac{\partial}{\partial r}\right) - \frac{Ze^2K}{r} \right| \psi_{1s} \right\rangle = E$$

$$\left\langle \psi_{1s} \left| \frac{-\hbar^2}{2\mu r^2}\left(\frac{\partial}{\partial r}r^2\frac{\partial}{\partial r}\right) \right| \psi_{1s} \right\rangle + \left\langle \psi_{1s} \left| \frac{-Ze^2K}{r} \right| \psi_{1s} \right\rangle = E.$$

The value of the potential energy, E_{pot}, is found using equation (2.40):

$$\left\langle \psi_{1s} \left| \frac{-Ze^2K}{r} \right| \psi_{1s} \right\rangle = -Ze^2K \times \frac{Z}{a_0} = -\frac{Z^2e^2K}{a_0}.$$

For the more complicated case of the kinetic energy expression, it can be instructive to investigate the result of the radial operator acting on the ket radial wavefunction in isolation, omitting multiplicative constants (since these are only reproduced unchanged). After the first differentiation and collecting constants outside the differential term, the following result is obtained:

$$\frac{\hbar^2\rho}{4\mu r^2}\frac{\partial}{\partial r}(r^2\mathrm{e}^{-\rho r/2}) = E_{\mathrm{kin}}\mathrm{e}^{-\rho r/2}$$

$$\frac{\hbar^2\rho}{4\mu r^2}\left(2r - \frac{\rho r^2}{2}\right)\mathrm{e}^{-\rho r/2} = E_{\mathrm{kin}}\mathrm{e}^{-\rho r/2}.$$

The above expression for kinetic energy factorises into two parts. The r^2 in the denominator of the first fraction cancels the r^2 in the numerator of the second term in the brackets, leaving the constant term $-\frac{\hbar^2\rho^2}{8\mu}$. Combining equations (2.26) and (2.27) gives an expression for ρ in terms of the same constants found in the expression for the potential energy:

$$\rho = \frac{2Z\mu e^2 K}{n\hbar^2}.$$

If we substitute this expression into the constant above and substitute in a_0 using equation (2.26) we get a useful result:

$$-\frac{\hbar^2\rho^2}{8\mu} = -\frac{Z^2e^4K^2\mu}{2\hbar^2n^2} = -\frac{Z^2e^2K}{2a_0}.$$

This second part need not be multiplied by the bra wavefunction and integrated, as the process will only return the same constant for that component of the kinetic energy. The first term in the brackets will however remain a function of r and so will need to be multiplied by the square of the wavefunction and integrated over all r using the Jacobian r^2. The integral is solved by parts:

$$\begin{aligned}\int_0^\infty \left\{\left(\frac{Z}{a_0}\right)^{\frac{3}{2}} 2\mathrm{e}^{-\rho r/2}\right\}^2 \frac{\hbar^2\rho}{8\mu r} r^2\,\mathrm{d}r &= \frac{\hbar^2Z^3\rho}{2\mu a_0^3}\int_0^\infty \frac{\mathrm{e}^{-\rho r}}{r} r^2\,\mathrm{d}r \\ &= \frac{\hbar^2Z^3\rho}{2\mu a_0^3}\left[-\mathrm{e}^{-\rho r}\left(\frac{r}{\rho} + \frac{1}{\rho^2}\right)\right]_0^\infty\end{aligned}$$

$$= \frac{\hbar^2 Z^3 \rho}{2\mu a_0^3} \cdot \frac{1}{\rho^2}$$

$$= \frac{\hbar^2 Z^3}{2\mu \rho a_0^3}.$$

If we substitute equation (2.26) for one of the Bohr radii above and equation (2.27) for ρ, we arrive at $\frac{Z^2 e^2 K}{a_0}$. This term cancels out the potential energy term, while the other kinetic energy term is half the potential energy, which thus remains as the total energy. This is in accordance with the virial theorem and with equation (2.22). Having solved this equation, it is evident that the 1s wavefunction is a solution to the Schrödinger equation, which was also required.

Exercise 22 We need to show that $\langle \psi_{1s} | \psi_{2s} \rangle = 0$. To integrate these radial functions over all space we need to use the Jacobian r^2. Since we are looking for the integral to cancel to zero we can ignore the constants multiplying the radial functions. We substitute in the radial exponential functions from Table 2.8. We must expand ρ as it depends on n so will be different for the two functions:

$$\left\langle e^{-\frac{Zr}{a_0}} \middle| \left(2 - \frac{Z}{a_0} r\right) e^{-\frac{Zr}{2a_0}} \right\rangle = \int_0^\infty \left(2e^{-\frac{3Zr}{2a_0}} - \frac{Z}{a_0} r e^{-\frac{3Zr}{2a_0}}\right) r^2 \, dr.$$

We need to apply the integral result of equation (A.15) and to extend it to the analogous result for x^3:

$$\int x^3 e^{-kx} \, dx = -e^{-kx} \left(\frac{x^3}{k} + \frac{3x^2}{k^2} + \frac{6x}{k^3} + \frac{6}{k^4}\right) + c. \tag{A.16}$$

We can now solve the integral. Let $k = \frac{3Z}{2a_0}$. Then

$$\left[-2e^{-kr}\left(\frac{r^2}{k} + \frac{2r}{k^2} + \frac{2}{k^3}\right) + \frac{2k}{3} e^{-kr}\left(\frac{r^3}{k} + \frac{3r^2}{k^2} + \frac{6r}{k^3} + \frac{6}{k^4}\right)\right]_0^\infty = 0.$$

This result confirms that ψ_{1s} and ψ_{2s} are orthogonal.

Exercise 23 In a one-electron atom, the most probable distance of the electron from the nucleus is the maximum in its radial probability density function, $4\pi r^2 \psi^2$. We insert $\psi_{1s} = 2(Z/a_0)^{3/2} e^{-\rho r/2}$, differentiate with respect to r, set to zero and solve for r:

$$16\pi \left(\frac{Z}{a_0}\right)^3 \frac{\mathrm{d}}{\mathrm{d}r}\left(r^2 \mathrm{e}^{-\rho r}\right) = 0$$

$$16\pi \left(\frac{Z}{a_0}\right)^3 \mathrm{e}^{-\rho r}\left(2r - \rho r^2\right) = 0$$

$$2r - \rho r^2 = 0$$

$$r = \frac{2}{\rho}.$$

Substituting in equation (2.27) with $Z = n = 1$ gives $r = a_0$ as desired.

Exercise 24 Radial nodes occur where $\psi(r) = 0$. For $\psi(r)_{2s}$, the function is zero when $2 - \rho r = 0$. This looks like the solution to the maximum for the 1s probability density function, except ρ, which depends on n, will be different. Substituting in equation (2.27) with $Z = 1$ and $n = 2$ gives $r = 2a_0$ for the radial node, as desired.

Exercise 25 Equation (2.31) contains the product of two terms involving r, so when differentiated with the product rule it will produce a sum of two terms which will need to be set to zero. Hence we can ignore the constant term and focus on the terms in r:

$$\frac{\mathrm{d}}{\mathrm{d}r}\left\{r^{2n} \exp\left(\frac{-2Zr}{na_0}\right)\right\} = \exp\left(\frac{-2Zr}{na_0}\right)\left(2nr^{2n-1} - \frac{2Z}{na_0}r^{2n}\right) = 0$$

$$r = \frac{n^2 a_0}{Z}.$$

This shows that equation (2.32) is correct.

Exercise 26 We use equation (2.34) and find the energy difference between the $n = 1$ and $n = 2$ levels:

$$\Delta E = -Z^2 R_H \left(\frac{1}{2^2} - \frac{1}{1^2}\right) = \frac{3R_H}{4}.$$

Given that $R_H = 1312\,\mathrm{kJ\ mol^{-1}}$, $\Delta E = 984\,\mathrm{kJ\ mol^{-1}}$. The ionisation energy of hydrogen is the Rydberg energy, i.e. $1312\,\mathrm{kJ\ mol^{-1}}$.

Exercise 27 We combine $c = f\lambda$ and equation (1.9) to find the ΔE associated with 1.21 nm and equate this to the Rydberg expression, equation (2.34), for $n = 2$ to $n = 1$. We need to express R_H as 2.18×10^{-18} J.

$$\frac{hc}{1.21 \times 10^{-9}\,\mathrm{m}} = -Z^2 R_H \left(\frac{1}{2^2} - \frac{1}{1^2}\right)$$

$$Z^2 = \frac{4hc}{3(2.18 \times 10^{-18}\,\mathrm{J})(1.21 \times 10^{-9}\,\mathrm{m})}$$

$$Z = 10.0$$

The element responsible was neon.

Exercise 28 We need to show that $\left\langle \psi_{1s} \,\middle|\, \frac{1}{r} \,\middle|\, \psi_{1s} \right\rangle = \frac{Z}{n^2 a_0}$. We substitute in the radial function from Table 2.8 and apply the Jacobian r^2 to the integration:

$$4\left(\frac{Z}{a_0}\right)^3 \int_0^\infty \frac{\mathrm{e}^{-\rho r}}{r} r^2\,\mathrm{d}r.$$

We proceed by integrating by parts:

$$4\left(\frac{Z}{a_0}\right)^3 \left[-\mathrm{e}^{-\rho r}\left(\frac{r}{\rho} + \frac{1}{\rho^2}\right)\right]_0^\infty = \frac{4}{\rho^2}\left(\frac{Z}{a_0}\right)^3.$$

We substitute in equation (2.27) to reach the answer, $\frac{Z}{n^2 a_0}$.

Exercise 29 We multiply $\left\langle \psi \,\middle|\, \frac{1}{r} \,\middle|\, \psi \right\rangle$ by $-Ze^2K$ to obtain the potential energy. Using the virial theorem (equation (2.41)) we subtract half the potential energy to arrive at the total energy, which is consistent with equation (2.33).

Exercise 30 There is no difference between the radial functions of each of the three 2p orbitals, and so the radial part of the Schrödinger equation will give the same energy to them all. We can therefore limit our comparison to the angular functions and the angular operators of equation (2.22).

It is convenient to calculate the energies from the angular functions in units of $\frac{\hbar^2}{2\mu r^2}$. Since this 'unit' contains the variable r, it might said that we need to multiply this by the square of the (real) wavefunction and integrate over all space. However, this integration involves the Jacobian r^2 which will cancel the r^2 in the denominator of our 'unit', leaving a constant. Since the wavefunctions are normalised, the integration will return the same result except with the variable r replaced by the constant a_0.

We begin by seeing the effect of the operators on the trigonometric wavefunctions without their normalisation constants or any integration. If they return the wavefunction multiplied by a constant we can use this constant without the need to use a bra wavefunction and integrate.

First we consider the p_z orbital. Since it is not a function of φ we can omit the differential operator with respect to this variable. The differential operator in θ gives the following result:

$$\begin{aligned} -\frac{1}{\sin\theta}\frac{\partial}{\partial\theta}\sin\theta\frac{\partial}{\partial\theta}(\cos\theta) &= E_{2\mathrm{p}_z}\cos\theta \\ 2\cos\theta &= E_{2\mathrm{p}_z}\cos\theta \\ 2 &= E_{2\mathrm{p}_z}. \end{aligned}$$

The p_x and p_y orbitals are functions of both θ and φ, so we need both sets of operator. First we consider the p_x orbital:

$$\begin{aligned} \left(-\frac{1}{\sin\theta}\frac{\partial}{\partial\theta}\sin\theta\frac{\partial}{\partial\theta} - \frac{1}{\sin^2\theta}\frac{\partial^2}{\partial\varphi^2}\right)(\sin\theta\cos\varphi) &= E_{2\mathrm{p}_x}\sin\theta\cos\varphi \\ \frac{\sin^2\theta-\cos^2\theta}{\sin\theta}.\cos\varphi + \frac{\sin\theta\cos\varphi}{\sin^2\theta} &= E_{2\mathrm{p}_x}\sin\theta\cos\varphi \\ \frac{2\sin^2\theta-1}{\sin\theta}.\cos\varphi + \frac{1}{\sin\theta}.\cos\varphi &= E_{2\mathrm{p}_x}\sin\theta\cos\varphi \\ 2\sin\theta\cos\varphi - \frac{\cos\varphi}{\sin\theta} + \frac{\cos\varphi}{\sin\theta} &= E_{2\mathrm{p}_x}\sin\theta\cos\varphi. \end{aligned}$$

Hence $E_{2\mathrm{p}_x} = 2$. Next we consider the p_y orbital:

$$\begin{aligned} \left(-\frac{1}{\sin\theta}\frac{\partial}{\partial\theta}\sin\theta\frac{\partial}{\partial\theta} - \frac{1}{\sin^2\theta}\frac{\partial^2}{\partial\varphi^2}\right)(\sin\theta\sin\varphi) &= E_{2\mathrm{p}_y}\sin\theta\sin\varphi \\ \frac{\sin^2\theta-\cos^2\theta}{\sin\theta}.\sin\varphi + \frac{\sin\theta\sin\varphi}{\sin^2\theta} &= E_{2\mathrm{p}_y}\sin\theta\sin\varphi \\ \frac{2\sin^2\theta-1}{\sin\theta}.\sin\varphi + \frac{1}{\sin\theta}.\sin\varphi &= E_{2\mathrm{p}_y}\sin\theta\sin\varphi \\ 2\sin\theta\sin\varphi - \frac{\sin\varphi}{\sin\theta} + \frac{\sin\varphi}{\sin\theta} &= E_{2\mathrm{p}_y}\sin\theta\sin\varphi. \end{aligned}$$

Hence $E_{2\mathrm{p}_y} = 2$. We have shown that the energy of all three 2p orbitals is the same.

Exercise 31 Since electrons in these two subshells have the same n, they must have the same potential energy, as made clear by equation (2.40). Therefore we need to show that the kinetic energies of the 2s and 2p electrons are equal.

The kinetic energy of the 2s electron, $E_{\mathrm{kin(2s)}}$, is only linear and so is found from the first term in equation (2.22). It will be instructive to see how the operator acts on the exponential term of the 2s radial wavefunction. We can omit the multiplicative factors of the exponential functions as they will cancel out.

$$\frac{-\hbar^2}{2\mu r^2}\frac{\partial}{\partial r}r^2\frac{\partial}{\partial r}\left\{(2-\rho r)\mathrm{e}^{-\rho r/2}\right\} = E_{\mathrm{kin(2s)}}(2-\rho r)\mathrm{e}^{-\rho r/2}$$

$$\frac{-\hbar^2}{2\mu r^2}\frac{\partial}{\partial r}\left\{\mathrm{e}^{-\rho r/2}\left(-2\rho r^2 + \frac{\rho^2 r^3}{2}\right)\right\} = E_{\mathrm{kin(2s)}}(2-\rho r)\mathrm{e}^{-\rho r/2}$$

$$\frac{-\hbar^2}{2\mu r^2}\mathrm{e}^{-\rho r/2}\left(-4\rho r + \frac{5\rho^2 r^2}{2} - \frac{\rho^3 r^3}{4}\right) = E_{\mathrm{kin(2s)}}(2-\rho r)\mathrm{e}^{-\rho r/2}$$

$$\frac{-\hbar^2}{2\mu r^2}\left(-2\rho r + \frac{\rho^2 r^2}{4}\right)(2-\rho r)\mathrm{e}^{-\rho r/2} = E_{\mathrm{kin(2s)}}(2-\rho r)\mathrm{e}^{-\rho r/2}$$

This expression can factorise according to the two terms in the first brackets. In the second term, the r^2 in the numerator cancels the r^2 in the denominator of the first fraction, leaving the constant $\frac{-\hbar^2\rho^2}{8\mu}$. This will be unaffected by multiplication by the bra wavefunction and integrating. The first term, however, remains a function of r. It is therefore multiplied by the square of the wavefunction and integrated over all r using the Jacobian r^2:

$$\int_0^\infty \frac{\hbar^2\rho}{\mu r}\left(\frac{Z^3}{8a_0^3}(2-\rho r)^2\mathrm{e}^{-\rho r}\right) r^2\,\mathrm{d}r$$

$$= \frac{\hbar^2 Z^3\rho}{8\mu a_0^3}\int_0^\infty \frac{4-4\rho r+\rho^2 r^2}{r}\mathrm{e}^{-\rho r}\,r^2\,\mathrm{d}r.$$

This integral factorises into three, each of which is solved by parts. Since the integration is between 0 and ∞, the only non-zero resultants will come from the terms in ρ that do not involve r. Only these terms are shown (in the inner brackets):

$$\frac{\hbar^2 Z^3 \rho}{8\mu a_0^3}\left[-\mathrm{e}^{-\rho r}\left(\frac{4}{\rho^2}-\frac{8}{\rho^2}+\frac{6}{\rho^2}\right)\right]_0^\infty = \frac{\hbar^2 Z^3}{4\mu a_0^3 \rho}.$$

For the 2p electron we need to calculate rotational as well as linear kinetic energy. We choose the $2p_z$ orbital for simplicity. We apply the method above to the linear kinetic energy:

$$\frac{-\hbar^2}{2\mu r^2}\frac{\partial}{\partial r}r^2\frac{\partial}{\partial r}\left(\rho r \mathrm{e}^{-\rho r/2}\right) = E_{\mathrm{kin,lin(2p)}}\rho r \mathrm{e}^{-\rho r/2}$$

$$\frac{-\hbar^2}{2\mu r^2}\frac{\partial}{\partial r}\left\{\mathrm{e}^{-\rho r/2}\left(\rho r^2 - \frac{\rho^2 r^3}{2}\right)\right\} = E_{\mathrm{kin,lin(2p)}}\rho r \mathrm{e}^{-\rho r/2}$$

$$\frac{-\hbar^2}{2\mu r^2}\left(2 - 2\rho r + \frac{\rho^2 r^2}{4}\right)\rho r \mathrm{e}^{-\rho r/2} = E_{\mathrm{kin,lin(2p)}}\rho r \mathrm{e}^{-\rho r/2}.$$

This expression factorises according to the three terms in the brackets. The second and third of these are identical to what was obtained for the kinetic energy of the 2s electron. The r^2 in the numerator of the third term cancels the r^2 in the denominator of the first fraction, leaving the same constant term that was found for the 2s kinetic energy. This constant will be unaffected by the multiplication by the bra wavefunction and the integration. The second term in the bracket will remain a function of r but will produce the same result when the full integration is carried out. This is shown below:

$$\begin{aligned}
&\frac{\hbar^2 \rho}{\mu}\int_0^\infty \frac{1}{r}\left\{\frac{1}{2\sqrt{6}}\left(\frac{Z}{a_0}\right)^{\frac{3}{2}}\rho r \mathrm{e}^{-\rho r/2}\right\}^2 r^2\,\mathrm{d}r \\
&= \frac{\hbar^2 Z^3 \rho^3}{24\mu a_0^3}\int_0^\infty r\mathrm{e}^{-\rho r}\, r^2\,\mathrm{d}r \\
&= \frac{\hbar^2 Z^3 \rho^3}{24\mu a_0^3}\left[-\mathrm{e}^{-\rho r}\left(\frac{r^3}{\rho}+\frac{3r^2}{\rho^2}+\frac{6r}{\rho^3}+\frac{6}{\rho^4}\right)\right]_0^\infty \\
&= \frac{\hbar^2 Z^3}{4\mu a_0^3 \rho}.
\end{aligned}$$

For the rotational kinetic energy, $E_{\mathrm{kin,rot(2p)}}$, of a $2p_z$ electron we can omit the constants that multiply the function of θ and just apply the first of the rotational kinetic energy operators (as the second only operates on φ):

$$\frac{-\hbar^2}{2\mu r^2}\frac{1}{\sin\theta}\frac{\partial}{\partial\theta}\sin\theta\frac{\partial}{\partial\theta}(\cos\theta) = E_{\text{kin,rot(2p)}}\cos\theta$$

$$\frac{-\hbar^2}{2\mu r^2}\cdot(-2)\cos\theta = E_{\text{kin,rot(2p)}}\cos\theta.$$

This result is a function of r and so would require multiplication by the square of the radial function and integration. However, it is identical except for a minus sign to the function of r that was the first term from the linear kinetic energy of the 2p electron. Hence the two terms cancel, leaving the overall kinetic energy for the 2p electron identical to that of the 2s electron. Since their potential energies are also the same, this demonstrates that the 2s and 2p electrons in hydrogen have the same energy, as desired.

Exercise 32 We relate the velocity operator to the momentum operator and the Laplacian:

$$\hat{v} = \frac{\hat{p}}{m} = \frac{\hbar}{im}\nabla.$$

We now need to express the Laplacian in spherical polar coordinates. Since the 1s orbital wavefunction does not depend on θ or φ, we only need consider the radial part of the differential operator:

$$\nabla_r = \frac{\partial r}{\partial x}\times\frac{\partial}{\partial r}+\frac{\partial r}{\partial y}\times\frac{\partial}{\partial r}+\frac{\partial r}{\partial z}\times\frac{\partial}{\partial r}.$$

Substituting in the differentials from equation (A.3) gives the laplacian for radial coordinates.

$$\nabla_r = \{\sin\theta(\cos\varphi+\sin\varphi)+\cos\theta\}\frac{\partial}{\partial r}$$

Differentiation of the 1s wavefunction with respect to r returns the wavefunction multiplied by $-\rho/2$ or $-Z/a_0$. To find the velocity expectation value we now need to multiply by the bra wavefunction and integrate over all space, following equation (2.39). Since the 1s wavefunction is normalised, its square multiplied by the Jacobian r^2 and integrated from $r = 0$ to ∞ will give 1. The functions of φ will integrate between $\varphi = 0$ and 2π to give zero. The functions of θ need to be multiplied by the Jacobian $\sin\theta$ and then integrated between $\theta = 0$ and π. Since $\cos\theta\sin\theta\,\mathrm{d}\theta$ integrates to $\frac{1}{2}\sin^2\theta$, which between the limits of 0 and π returns zero, all the terms in the integral to find the velocity expectation value come to zero, demonstrating that the expectation value of the velocity is zero.

Exercise 33 The virial theorem tells us that $E_{\text{pot}} = -2E_{\text{kin}}$. Since $E_{\text{total}} = E_{\text{pot}} + E_{\text{kin}}$, it follows that $E_{\text{kin}} = -E_{\text{total}}$. We insert the values of the constants into minus equation (2.33) and multiply by the Avogadro constant, N_A:

$$\begin{aligned} E_{\text{kin}} &= \frac{Z^2 e^2 K N_A}{2n^2 a_0} \\ &= \frac{(1.6022 \times 10^{-19}\,\text{C})^2 \times 8.9876 \times 10^9\,\text{J C}^{-2}\,\text{m} \times 6.0221 \times 10^{23}\,\text{mol}^{-1}}{2 \times 5.2918 \times 10^{-11}\,\text{m}} \\ &= 1.313 \times 10^6\,\text{J mol}^{-1}. \end{aligned}$$

This shows a remarkable agreement between experiment and theory, considering the Schrödinger equation adopted does not take into account special relativity. It is worth bearing in mind that the value of the Bohr radius used here is the conventional one using the mass of the electron rather than the reduced mass, which makes the Bohr radius slightly larger and the ionisation energy slightly smaller. Using the reduced mass of the electron, the Bohr radius for the hydrogen atom comes to 5.2947×10^{-11} m, which gives a kinetic energy of 1312 kJ mol^{-1}, as observed.

Exercise 34 The virial theorem tells us that E_{kin} equals the ionisation energy. We divide by the Avogadro constant to give the ionisation energy of a hydrogen 1s electron as 2.18×10^{-18} J:

$$v = \sqrt{\frac{2E_{\text{kin}}}{m}} = \sqrt{\frac{2 \times 2.18 \times 10^{-18}\,\text{J}}{9.11 \times 10^{-31}\,\text{kg}}} = 2.2 \times 10^6\,\text{m s}^{-1}.$$

Dividing by $3.00 \times 10^8\,\text{m s}^{-1}$ gives the speed of the electron as 0.73% of the speed of light.

Chapter 3: Multi-Electron Atoms

Exercise 35 To find the ionisation energy we need to sum all the electron energies in the oxygen atom and subtract this from the sum of the electron energies in the O^+ ion. We can neglect the energies of the inner electrons from this calculation as their contribution will cancel out. In the oxygen atom each valence electron experiences five shielding contributions of 0.35

from the other valence electrons and 2×0.85 from the inner electrons, giving a total shielding of $\sigma = 3.45$. Given that $Z = 8$ for oxygen, this gives $Z_{\text{eff}} = 4.55$. Substituting Z_{eff} for Z in equation (2.34) and multiplying the energy by six valence electrons gives the total valence electron energy as $E_{\text{O(val)}} = -31.05R_{\infty}$. R_{∞} is based on the rest mass of the electron rather than its reduced mass in hydrogen since oxygen is significantly heavier than hydrogen. The same process for the five valence electrons of O^+ gives the result $E_{\text{O+(val)}} = -30.01R_{\infty}$. Given that R_{∞}=1312.7 kJ mol^{-1}, the difference between the two valence electron energies comes to 1367 kJ mol^{-1}. This isn't a bad approximation of the experimental value of 1314 kJ mol^{-1} [39].

Exercise 36 It is convenient first to convert the molar ionisation energy into units of R_{∞} (again using the rest mass of the electron rather than its reduced mass in hydrogen). Dividing the ionisation energy by 1312.7 kJ mol$^{-1}/R_{\infty}$ gives the ionisation energy as $1.807\,R_{\infty}$. This ionisation energy is the difference between the energy of the sole electron in He^+ and the sum of the energies of the two electrons in He. Since the electron in He^+ evidently experiences no shielding, its energy from equation (2.34) must be $-4R_{\infty}$. Hence the energy of each electron in He must be $\frac{1}{2}\left(-4\,R_{\infty} - 1.807\,R_{\infty}\right) = -2.903\,R_{\infty}$. This gives $Z_{\text{eff}} = \sqrt{2.903} = 1.70$. Therefore the shielding factor, $\sigma = Z - 1.70 = 0.30$, as was intended.

Exercise 37 Like the previous exercise, we convert the molar ionisation energy into units of R_{∞}, giving $0.396\,R_{\infty}$. Since lithium only has one valence electron, the energy of this electron is equal to minus the ionisation energy (as the energies of the core electrons are not affected by the valence electron in Slater's rules). Using equation (2.34), $Z_{\text{eff}} = \sqrt{0.396 \times 2^2} = 1.26$. Therefore the shielding factor for each core electron, $\sigma = \frac{1}{2}(Z - 1.26) = 0.87$, which is close to the 0.85 that is used. 0.85 would have been chosen to give a satisfactory result over a broad range of elements.

Exercise 38 Again, we convert the molar ionisation energy into units of R_{∞}, giving $1.001\,R_{\infty}$. This is the difference between the total energy of the valence electrons in O^+ and the total energy of the valence electrons in O. The core electrons contribute $\sigma = 2 \times 0.85 = 1.70$ to the shielding of each valence electron. Each valence electron in O^+ has $Z_{\text{eff}} = 8 - (1.70 + 4\sigma)$ where σ is the shielding contribution of one valence electron felt by another.

Each valence electron in O has $Z_{\text{eff}} = 8 - (1.70 + 5\sigma)$. We find σ by using the total valence energies and equation (2.34): $E_{\text{O(val)}} + \text{IE(O)} = E_{\text{O}^+\text{(val)}}$.

$$6\left(\frac{-(6.30 - 5\sigma)^2 R_\infty}{2^2}\right) + 1.001 R_\infty = 5\left(\frac{-(6.30 - 4\sigma)^2 R_\infty}{2^2}\right)$$

$$24(6.3 - 5\sigma)^2 - 20(6.3 - 4\sigma)^2 = 4.004$$

$$280\sigma^2 - 504\sigma + 154.756 = 0$$

$$\sigma = 0.39, 1.41$$

The solution $\sigma = 1.41$ is not viable as it cannot exceed 1. However, $\sigma = 0.39$ is a realistic solution as 0.35 is a value chosen to give a satisfactory result over a broad range of elements. It certainly seems a better value than 0.30 for oxygen.

Exercise 39

$$\text{Li} : Z_{\text{eff}} = 3 - (2 \times 0.85) = 1.30 \quad \text{IE} = 1.3^2 R_\infty/2^2 = 555 \text{ kJ mol}^{-1}$$

$$\text{Na} : Z_{\text{eff}} = 11 - (6.8 + 2) = 2.20 \quad \text{IE} = 2.2^2 R_\infty/3^2 = 706 \text{ kJ mol}^{-1}$$

$$\text{K} : Z_{\text{eff}} = 19 - (6.8 + 10) = 2.20 \quad \text{IE} = 2.2^2 R_\infty/3.7^2 = 464 \text{ kJ mol}^{-1}$$

$$\text{Rb} : Z_{\text{eff}} = 37 - (6.8 + 28) = 2.20 \quad \text{IE} = 2.2^2 R_\infty/4.0^2 = 397 \text{ kJ mol}^{-1}$$

$$\text{Cs} : Z_{\text{eff}} = 55 - (6.8 + 46) = 2.20 \quad \text{IE} = 2.2^2 R_\infty/4.2^2 = 360 \text{ kJ mol}^{-1}$$

$$\text{Fr} : Z_{\text{eff}} = 87 - (6.8 + 78) = 2.20 \quad \text{IE} = 2.2^2 R_\infty/4.3^2 = 344 \text{ kJ mol}^{-1}$$

The most important differences between the Slater-predicted ionisation energies and the experimental ones occur for lithium and francium, which are out of sequence. Lithium is the odd one out: in all the others there are eight imperfectly shielding electrons in the shell beneath the valence shell, providing the contribution of 6.8 towards σ. Lithium only has two such electrons and so is better shielded according to Slater's rules, and hence it is predicted to have a lower effective nuclear charge than the other alkali metals. Francium has a higher ionisation energy than caesium on account of the relativistic contraction of its 7s orbital (see Section 3.10).

Exercise 40 Substituting the values into the equation gives $\chi = 3.06$.

Exercise 41 A shell with quantum number n contains n subshells, whose quantum number l begins at 0. Each subshell holds $2l+1$ orbitals, which can each hold two electrons. We therefore need to calculate the sum

$$\sum_{l=0}^{l=n-1} 2(2l+1) = 2n^2.$$

Exercise 42 To find the configuration-average energy, the energy of each level needs to be multiplied by the degeneracy of the level. The degeneracy, g, of each level and its product with the energy are shown in Table A.1. The sum of these energies over the configuration then needs to be divided by the total degeneracy of the configuration. The degeneracy of $4s^2 3d^1$ is $1 \times 10 = 10$, and for $4s^1 3d^2$ it is $2 \times \frac{10!}{2!\ 8!} = 90$. This gives the average energy for $4s^2 3d^1$ as 1008 cm^{-1}/10 = 101 cm^{-1} and $4s^1 3d^2$ as 1454868 cm^{-1}/90 = 16165 cm^{-1}. The difference in the configuration energies is therefore 16064 cm^{-1}, which is $(16064 \times 1.240 \times 10^{-4})$ eV = 1.99 eV.

Table A.1. Calculation of the $4s^2 3d^1$ and $4s^1 3d^2$ relative configurations-average energy in Sc.

Config.	Term	J	E/cm^{-1}	g	$E \times g$
$4s^2 3d^1$	^{2}D	1.5	0	4	0
$4s^2 3d^1$	^{2}D	2.5	168	6	1008
$4s^1 3d^2$	^{4}F	1.5	11520	4	46080
$4s^1 3d^2$	^{4}F	2.5	11558	6	69348
$4s^1 3d^2$	^{4}F	3.5	11610	8	92880
$4s^1 3d^2$	^{4}F	4.5	11677	10	116770
$4s^1 3d^2$	^{2}F	2.5	14926	6	89556
$4s^1 3d^2$	^{2}F	3.5	15042	8	120336
$4s^1 3d^2$	^{2}D	2.5	17013	6	102078
$4s^1 3d^2$	^{2}D	1.5	17025	4	68100
$4s^1 3d^2$	^{4}P	0.5	17226	2	34452
$4s^1 3d^2$	^{4}P	1.5	17255	4	69020
$4s^1 3d^2$	^{4}P	2.5	17307	6	103842
$4s^1 3d^2$	^{2}G	4.5	20237	10	202370
$4s^1 3d^2$	^{2}G	3.5	20240	8	161920
$4s^1 3d^2$	^{2}P	0.5	20681	2	41362
$4s^1 3d^2$	^{2}P	1.5	20720	4	82880
$4s^1 3d^2$	^{2}S	0.5	26937	2	53874

Exercise 43 A useful result that we apply to this question is the following general integral result:

$$\int_0^\infty r^n e^{-kr}\, r^2\, dr = \frac{1}{k^{n+3}} \frac{d^{n+2}}{dr^{n+2}}(r^{n+2}). \tag{A.17}$$

When the integral $\langle \psi(1s)|\psi(2s)\rangle$ is set to zero, it is clear from equation (A.16) and the following step that the integral simplifies to

$$2\left(\frac{2}{k^3}\right) = \frac{2k}{3}\left(\frac{6}{k^4}\right),$$

which is obviously true. However, the two k's that appear on the right-hand side of this expression have different origins: the first is from the pure 2s function, while the second is from the product of the 1s and 2s functions. The k on the left-hand side of the equation is also from the product of the 1s and 2s functions. If the 1s and 2s functions had different Bohr radii they would generate different k constants, and the equation above would no longer be true. This would leave the two wavefunctions not orthogonal to one another, but the orthogonality of eigenfunctions is a requirement (see Section 1.4). Therefore the Bohr radius must be the same for the two orbitals. This result can be generalised to the integral of any other s function with 1s. There will always be two types of constant, ρ: one from an individual orbital and the other from the product of two orbitals. Unless they are identical, i.e. they are built from the same Bohr radius, the orthogonality of the two orbitals can never be satisfied. Hence the reduced 1s Bohr radius imposed by special relativity must be replicated in the other s orbital radial functions.

Exercise 44 The s subshell has orbital angular momentum quantum number $l = 0$, so the associated Legendre function that describes its angular dependence is trivially 1 (see Table 2.1). However, the other subshells with non-zero l have associated Legendre functions that are orthogonal to 1, i.e. when multiplied by the Jacobian $\sin\theta$ and integrated between 0 and π the result is zero. Again, this is a requirement for the angular functions to be eigenfunctions. The other subshells are therefore always orthogonal to 1s regardless of the radial functions, so there is no requirement for the radial functions of the other subshells to change. Hence their Bohr radius is unaffected by the relativistic contraction of the 1s orbital.

Chapter 4: Wavefunctions in Multi-Electron Atoms

Exercise 45 There are $2l + 1$ orbitals in a subshell and therefore, considering spin, $4l + 2$ possible electron states within the subshell. Since each electron state may only be taken by one electron per atom, we need to find the number of ways to distribute n electrons among $4l + 2$ states:

$$^{4l+2}C_n = \frac{(4l+2)!}{n!(4l+2-n)!}.$$

Exercise 46 Subshells have the greatest degeneracy when they are half-full. The greatest degeneracy in the f subshell is therefore for f^7, commonly associated with Gd^{3+}. The degeneracy, g, is given by the following expression:

$$g = \frac{14!}{7!7!} = 3432.$$

Exercise 47 The number of possible microstates in p^3 is $^6C_3 = 20$. These microstates are described by M_S and M_L labels in Table A.2.

Table A.2. The microstates of p^3 in the Russell–Saunders coupling scheme.

e(1)	e(2)	e(3)	M_S	M_L
1^+	1^-	0^+	1/2	2
1^+	0^+	0^-	1/2	1
1^+	0^+	-1^-	1/2	0
0^+	0^-	-1^+	1/2	−1
0^+	-1^+	-1^-	1/2	−2
1^+	1^-	0^-	−1/2	2
1^-	0^+	0^-	−1/2	1
1^+	0^-	-1^-	−1/2	0
0^+	0^-	-1^-	−1/2	−1
0^-	-1^+	-1^-	−1/2	−2
1^+	1^-	-1^+	1/2	1
1^+	0^-	-1^+	1/2	0
1^+	-1^+	-1^-	1/2	−1
1^+	1^-	-1^-	−1/2	1
1^-	0^+	-1^-	−1/2	0
1^-	-1^+	-1^-	−1/2	−1
1^+	0^+	-1^+	3/2	0
1^-	0^+	-1^+	1/2	0
1^-	0^-	-1^+	−1/2	0
1^-	0^-	-1^-	−3/2	0

Exercise 48 Referring to Table A.2, we begin with the microstates with maximum M_L, which is 2. Those two microstates have $M_S = \pm 1/2$. Since Russell–Saunders terms are $(2S+1)(2L+1)$-fold degenerate, there will be ten of these microstates — the first ten in the table — with descending values of M_S and M_L making up the ^{2}D term. The next highest M_L value is 1. These two microstates have $M_S = \pm 1/2$ and so they belong to the six-fold degenerate ^{2}P Russell–Saunders term. These six microstates may be taken to be the ones beneath the top ten that were associated with the ^{2}D state. The four remaining microstates have $M_L = 0$ and $M_S = 3/2, 1/2, -1/2, -3/2$ so these belong to the ^{4}S term with $S = 3/2$ and $L = 0$. That takes account of the 20 possible microstates of p^3.

Exercise 49 The terms with the largest L in the f block will be associated with the microstates in which all positive m_l orbitals are doubly occupied and no negative m_l orbitals are occupied. These can be found in f^6, and also in f^7 and f^8 when additional electrons are in the $m_l = 0$ orbital. The total M_L in these cases will be 12, which is associated with the letter Q. In f^6 (Eu^{3+}) and f^8 (Tb^{3+}) this must be ^{1}Q since all the electrons must be paired, while in f^7 (Gd^{3+}) this must be ^{2}Q since there must be a single unpaired electron.

Exercise 50 Since each electron occupies a different p subshell, $6 \times 6 = 36$ microstates are possible. They are collected in Table A.3 together with their M_S and M_L values.

Without the Pauli constraint of p^2 we can maximise M_L while also maximising M_S, which gives a ^{3}D term; this accounts for the first 15 microstates. This leaves a microstate with $M_L = 2$ and $M_S = 0$, which is part of a ^{1}D term, accounting for the five microstates after the 15 ^{3}D microstates. This leaves a microstate with $M_L = 1$ and $M_S = 1$, which is part of a ^{3}P term, accounting for the next nine microstates. There remains a microstate with $M_L = 1$ and $M_S = 0$, which is part of a ^{1}P term, accounting for the next three microstates. This leaves four microstates all with $M_S = 0$. One has $M_L = 1$, which is part of ^{3}S, accounting for three of these four remaining microstates. The final one, with $M_L = 0$, is therefore the ^{1}S term.

Therefore the configuration pp generates the terms ^{3}D, ^{1}D, ^{3}P, ^{1}P, ^{3}S and ^{1}S, while p^2 generates just ^{1}D, ^{3}P and ^{1}S. It is a general feature of ll and l^2 configurations that L takes values in integer steps from $2L$ to 0. In ll configurations each value of L can be associated with a singlet and triplet term, while in l^2 configurations, the maximum L can only be a singlet,

Table A.3. The microstates of pp in the Russell–Saunders coupling scheme.

e(1)	e(2)	M_S	M_L
1^+	1^+	1	2
1^+	0^+	1	1
1^+	-1^+	1	0
0^+	-1^+	1	−1
-1^+	-1^+	1	−2
1^+	1^-	0	2
1^+	0^-	0	1
1^+	-1^-	0	0
0^+	-1^-	0	−1
-1^+	-1^-	0	−2
1^-	1^-	−1	2
1^-	0^-	−1	1
1^-	-1^-	−1	0
0^-	-1^-	−1	−1
-1^-	-1^-	−1	−2
1^-	1^+	0	2
1^-	0^+	0	1
1^-	-1^+	0	0
0^-	-1^+	0	−1
-1^-	-1^+	0	−2
0^+	1^+	1	1
0^+	0^+	1	0
-1^+	0^+	1	−1
0^+	1^-	0	1
0^+	0^-	0	0
-1^+	0^-	0	−1
0^-	1^-	−1	1
0^-	0^-	−1	0
-1^-	0^-	−1	−1
0^-	1^+	0	1
0^-	0^+	0	0
-1^-	0^+	0	−1
-1^+	1^+	1	0
-1^+	1^-	0	0
-1^-	1^-	−1	0
-1^-	1^+	0	0

and as L decreases in integer steps the spin multiplicity alternates between singlet and triplet. This restriction is imposed by the Pauli principle.

Exercise 51 The ll configurations have singlet and triplet terms up to $2L$. So dd has the terms ^{3}G, ^{1}G, ^{3}F, ^{1}F, ^{3}D, ^{1}D, ^{3}P, ^{1}P, ^{3}S and ^{1}S; ff has these terms and in addition ^{3}I, ^{1}I, ^{3}H and ^{1}H. In the l^2 configurations, the maximum L can only be a singlet, and as L decreases in integer steps the spin multiplicity alternates between singlet and triplet. So d^2 has the terms ^1G, ^3F, ^1D, ^3P and ^1S; f^2 has these terms and in addition ^1I and ^3H.

Exercise 52 In p^1 there are six microstates corresponding to the six possible states of an electron in a p subshell. These form a single term, ^2P. In p^6 there is only one microstate, where every possible state in the subshell is occupied. Summation of the m_s and m_l values therefore gives 0 for M_S and M_L, which is equivalent to p^0, i.e. no electrons and no angular momenta. Indeed there is a symmetry in the p subshell where p^n has the same terms as p^{6-n}, since a hole in a nearly full subshell configuration behaves rather like a positive electron would. Hence p^5 has just the one term like p^1, and p^4 has the same terms as p^2. The same symmetry is found in other subshells.

Exercise 53 Singlet states, i.e. those with $S = 0$, must have $J = L$, so their spin–orbit coupled levels are $^1D_2$ and $^1S_0$. Following equation (4.20), the ^3P term has levels $^3P_2$, $^3P_1$ and $^3P_0$.

Exercise 54 The degeneracy, g, of a J level is $2J + 1$, so $g(^3P_2) = 5$, $g(^3P_1) = 3$ and $g(^3P_0) = 1$. The sum of these degeneracies is 9, which was established as the degeneracy of the ^3P term, i.e. $(2S + 1)(2L + 1)$.

Exercise 55 The microstates of the ^2D term of p^3 and their M_J values are given in Table A.4.

Table A.4. The microstates of the ^2D term of p^3 including their M_J values.

e(1)	e(2)	e(3)	M_S	M_L	M_J
1^+	1^-	0^+	1/2	2	5/2
1^+	0^+	0^-	1/2	1	3/2
1^+	0^+	-1^-	1/2	0	1/2
0^+	0^-	-1^+	1/2	−1	−1/2
0^+	-1^+	-1^-	1/2	−2	−3/2
0^-	-1^+	-1^-	−1/2	−2	−5/2
1^+	1^-	0^-	−1/2	2	3/2
1^-	0^+	0^-	−1/2	1	1/2
1^+	0^-	-1^-	−1/2	0	−1/2
0^+	0^-	-1^-	−1/2	−1	−3/2

The sequence of M_J labels makes it clear that the first six might be assigned to $^2D_{5/2}$ while the remaining four might be assigned to $^2D_{3/2}$. This is consistent with the $(2J+1)$-fold degeneracy of J levels.

Exercise 56 Following Hund's rules, we maximise S and then maximise L. Due to electron-hole symmetry, the ground-state term for d^{10-n} will be the same as for d^n. The ground-state terms are collected in Table A.5.

Exercise 57 We use the ground-state terms, which will be equal for f^n and f^{14-n}, and select the lowest J level for the first half of the f^n series and the highest J level for the second half. The ground-state levels are collected in Table A.6.

Table A.5. The ground-state terms for d^1 to d^{10} in the Russell–Saunders coupling scheme.

Configuration	Ground-state term
d^1, d^9	^2D
d^2, d^8	^3F
d^3, d^7	^4F
d^4, d^6	^5D
d^5	^6S
d^{10}	^1S

Table A.6. The ground-state levels for f^1 to f^{14} in the Russell–Saunders coupling scheme.

Configuration	Ground-state level
f^1	$^2F_{5/2}$
f^2	$^3H_4$
f^3	$^4I_{9/2}$
f^4	$^5I_4$
f^5	$^6H_{5/2}$
f^6	$^7F_0$
f^7	$^8S_{7/2}$
f^8	$^7F_6$
f^9	$^6H_{15/2}$
f^{10}	$^5I_8$
f^{11}	$^4I_{15/2}$
f^{12}	$^3H_6$
f^{13}	$^2F_{7/2}$
f^{14}	$^1S_0$

Exercise 58 We use $(2J+1)$ to find the degeneracies of the $^3F_4$ ground state and $^3D_3$ excited state, giving $g_1 = 7$ and $g_0 = 9$. The equation is simplified by cancelling the factor 10^{-23} from the exponential:

$$\frac{n_1}{n_0} = \frac{7\exp\left(\frac{-205\times 1.986}{1.381\times 298}\right)}{9} = 0.372.$$

It follows that the percentage of gaseous nickel atoms in the ground state, $N_0 = (100 - N_0)/0.372$, and so $N_0(1 + \frac{1}{0.372}) = \frac{100}{0.372}$. Therefore $N_0 = \frac{100}{0.372} \times \frac{0.372}{1.372} = \frac{100}{1.372} = 73\%$.

Exercise 59 The microstates are collected in Table A.7. Note that Table A.7 is completed in the same systematic fashion as Table 4.3. The first row aims to minimise j, starting with j_1. Quantum numbers are then incremented, starting from the right-hand side. The first four microstates make up the $(\frac{1}{2}, \frac{1}{2}, \frac{3}{2})$ group, whose M values indicate that they belong to $J = \frac{3}{2}$ (note that the degeneracy of this grouping conforms to $2J+1$). The next 12 microstates make up the $(\frac{1}{2}, \frac{3}{2}, \frac{3}{2})$ group, which can be seen to belong to $J = \frac{5}{2}$, $\frac{3}{2}$ and $\frac{1}{2}$. These J levels have a total degeneracy of

Table A.7. The microstates of p^3 in the jj coupling scheme.

j_1	m_1	j_2	m_2	j_3	m_3	M
1/2	−1/2	1/2	1/2	3/2	−3/2	−3/2
					−1/2	−1/2
					1/2	1/2
					3/2	3/2
1/2	−1/2	3/2	−3/2	3/2	−1/2	−5/2
					1/2	−3/2
					3/2	−1/2
1/2	−1/2	3/2	−1/2	3/2	1/2	−1/2
					3/2	1/2
1/2	−1/2	3/2	1/2	3/2	3/2	3/2
1/2	1/2	3/2	−3/2	3/2	−1/2	−3/2
					1/2	−1/2
					3/2	1/2
1/2	1/2	3/2	−1/2	3/2	1/2	1/2
					3/2	3/2
1/2	1/2	3/2	1/2	3/2	3/2	5/2
3/2	−3/2	3/2	−1/2	3/2	1/2	−3/2
					3/2	−1/2
3/2	−3/2	3/2	1/2	3/2	3/2	1/2
3/2	−1/2	3/2	1/2	3/2	3/2	3/2

$6 + 4 + 2 = 12$, as is expected. The final four microstates make up the $(\frac{3}{2}, \frac{3}{2}, \frac{3}{2})$ group, which evidently belongs to $J = \frac{3}{2}$.

The jj levels of the p^3 configuration can therefore be summarised as $(\frac{1}{2}, \frac{1}{2}, \frac{3}{2})_{3/2}$, $(\frac{1}{2}, \frac{3}{2}, \frac{3}{2})_{5/2,\ 3/2,\ 1/2}$ and $(\frac{3}{2}, \frac{3}{2}, \frac{3}{2})_{3/2}$.

Exercise 60 ^{235}U will experience a hyperfine interaction, due to its nucleus having spin, while ^{238}U will not. While the electronic structure of these atoms will be very similar, the ^{235}U levels will be split (unless they have $J = 0$). If a sample of gas-phase uranium atoms were irradiated with a frequency of light just sufficient to ionise a higher ^{235}U hyperfine level but not the unsplit level in ^{238}U, then the ^{235}U atoms could be selectively ionised and removed from the sample by an electric field. Such a technique is known as AVLIS: atomic vapour laser isotope separation.

Exercise 61 We can use the analogy of subshells by considering the symmetry labels of g orbitals in cubic symmetry, which are collected in Table 4.5. Hence ^{1}G (degeneracy $g = 9$) splits into $^1A_{1g}(g = 1)$, $^1E_g(g = 2)$, $^1T_{1g}(g = 3)$ and $^1T_{2g}(g = 3)$. Summing the degeneracies for the crystal field levels gives 9, showing that the number of states of ^{1}G isn't changed by the crystal field.

Exercise 62 The Pr^{3+} f^2 configuration has a ground state of $^3H_4$ (see Table A.6). The $J = 4$ level is split like the $L = 4$ level, i.e. following the cubic g orbital labels, giving A_{1g}, E_g, T_{1g} and T_{2g}. Spin multiplicities are not attached to the labels as S is not a good quantum number after a spin–orbit coupling interaction.

Exercise 63 Extending the analogy to p^4, p^5 and p^6, we remember that p^n produces the same Russell–Saunders terms as p^{6-n}. Hence T^4, like T^2, has the crystal-field states $^3T_{1g}$, ^1E_g, $^1T_{2g}$ and $^1A_{1g}$. T^5 therefore is equivalent to T, and T^6, like p^0, is simply the zero angular momentum state ^{1}S and therefore $^1A_{1g}$ in an octahedral crystal field.

Exercise 64 We use p^3, which gives the Russell–Saunders terms ^{4}S, ^{2}D and ^{2}P. In an octahedral crystal field these terms become $^4A_{1g}$, ^2E_g, $^2T_{2g}$ and $^2T_{1g}$. The only difference from the actual crystal-field states is that this model gives $^4A_{1g}$ instead of the desired $^4A_{2g}$.

Chapter 5: Orbitals in Molecules

Exercise 65 Beginning with the bonding solution, we can insert the bonding energy into either secular equation. Using the first secular equation we obtain

$$c_1\left(\alpha - \frac{\alpha+\beta}{1+S}\right) + c_2\left(\beta - \frac{S(\alpha+\beta)}{1+S}\right) = 0$$

$$c_1\left(\frac{\alpha(1+S)-\alpha-\beta}{1+S}\right) + c_2\left(\frac{\beta(1+S)-S(\alpha+\beta)}{1+S}\right) = 0$$

$$c_1(\alpha S - \beta) + c_2(-\alpha S + \beta) = 0$$

$$c_1(\alpha S - \beta) - c_2(\alpha S - \beta) = 0$$

$$c_1 = c_2.$$

We can similarly insert the antibonding energy into the first secular equation:

$$c_1\left(\alpha - \frac{\alpha-\beta}{1-S}\right) + c_2\left(\beta - \frac{S(\alpha-\beta)}{1-S}\right) = 0$$

$$c_1(-\alpha S + \beta) + c_2(-\alpha S + \beta) = 0$$

$$c_1 = -c_2.$$

Exercise 66 To normalise an MO we integrate its square over all space, set the result to 1 and solve for c. There is no need for a complex conjugate as the atomic orbitals on which the MO are based are real. We neglect the cross terms involving atomic orbitals on different atoms, and use the fact that the atomic orbital wavefunctions are normalised. We get the following result using the bonding MO wavefunction:

$$\begin{aligned}
&\langle c(\psi_{1s}(1) + \psi_{1s}(2)) | c(\psi_{1s}(1) + \psi_{1s}(2)) \rangle \\
&= c^2\langle \psi_{1s}(1)|\psi_{1s}(1)\rangle + c^2\langle \psi_{1s}(2)|\psi_{1s}(2)\rangle \\
&= 2c^2 = 1 \\
c &= \frac{1}{\sqrt{2}}.
\end{aligned}$$

This gives the value for c that was desired. The same result is found using the antibonding MO wavefunction since the negative sign in front of $\psi_{1s}(2)$ returns a positive sign when it is squared.

Exercise 67 We expand out the determinant and then write it as a quadratic in E:

$$(\alpha_1 - E)(\alpha_2 - E) - \beta^2 = 0$$

$$E^2 - (\alpha_1 + \alpha_2)E + (\alpha_1\alpha_2 - \beta^2) = 0.$$

We solve the quadratic and express the square rooted term in the form $\sqrt{1+x}$:

$$E = \frac{(\alpha_1 + \alpha_2) \pm \sqrt{(\alpha_1 + \alpha_2)^2 - 4(\alpha_1\alpha_2 - \beta^2)}}{2}$$

$$E = \frac{(\alpha_1 + \alpha_2) \pm \sqrt{(\alpha_1 - \alpha_2)^2 + (2\beta)^2}}{2}$$

$$E = \frac{(\alpha_1 + \alpha_2) \pm (\alpha_1 - \alpha_2)\sqrt{1 + \frac{(2\beta)^2}{(\alpha_1 - \alpha_2)^2}}}{2}.$$

In a heteronuclear diatomic we assume that $(\alpha_1 - \alpha_2) \gg \beta$ and so we can proceed with the suggested approximation:

$$E \approx \frac{(\alpha_1 + \alpha_2) \pm (\alpha_1 - \alpha_2)\left(1 + \frac{1}{2}\frac{(2\beta)^2}{(\alpha_1 - \alpha_2)^2}\right)}{2}$$

$$E \approx \frac{(\alpha_1 + \alpha_2) \pm (\alpha_1 - \alpha_2)\left(1 + \frac{2\beta^2}{(\alpha_1 - \alpha_2)^2}\right)}{2}.$$

We express the larger root as E_1 and the smaller as E_2:

$$E_1 \approx \frac{(\alpha_1 + \alpha_2) + (\alpha_1 - \alpha_2)\left(1 + \frac{2\beta^2}{(\alpha_1 - \alpha_2)^2}\right)}{2}$$

$$E_1 \approx \alpha_1 + \frac{\beta^2}{(\alpha_1 - \alpha_2)^2}$$

$$E_2 \approx \frac{(\alpha_1 + \alpha_2) - (\alpha_1 - \alpha_2)\left(1 + \frac{2\beta^2}{(\alpha_1 - \alpha_2)^2}\right)}{2}$$

$$E_2 \approx \alpha_2 - \frac{\beta^2}{(\alpha_1 - \alpha_2)^2}.$$

Exercise 68 Since the s and three p orbitals are all mutually orthogonal, the product of the two hybrid orbitals will just involve the combination of like terms, leading to the squares of the functions. Since the atom orbitals are all real and normalised, their squares all equal to 1. This accounts for the integral coming to zero, as shown in the equation:

$$\langle \mathrm{sp}^2(1)|\mathrm{sp}^2(2)\rangle = \frac{1}{3}\left((-\mathrm{s})^2 - \mathrm{p}_x^2\right) = \frac{1}{3}(1-1) = 0.$$

Exercise 69 As hybrid orbitals are linear combinations of atomic orbital functions, the Hamiltonian operator is applied separately to each. Since the s orbital function is just 1 with no angular characteristics, it trivially returns an energy of zero to the angular part of the operator. The p_z function is $\cos\theta$. The term in the operator that is differential with respect to θ acts on $\cos\theta$ to return $\cos\theta$ multiplied by $\frac{\hbar^2}{\mu r^2}$, the latter being the angular energy, showing that p_z is also an eigenfunction of the Hamiltonian. Hence the sp(1) and sp(2) orbitals are both eigenfunctions of the Hamiltonian operator.

Exercise 70 We need to find N in $\psi_{MO} = N(\psi_{p_z}(1) + \psi_{p_z}(2) + \psi_{p_z}(3) + \psi_{p_z}(4) + \psi_{p_z}(5) + \psi_{p_z}(6))$. To normalise the MO we integrate its square over all space, set the result to 1 and solve for N. There is no need for a complex conjugate as the atomic orbitals on which the MO are based are real. We neglect the cross terms involving atomic orbitals on different atoms, and use the fact that the atomic orbital wavefunctions are normalised. For brevity, integrals of the type $\langle\psi_{p_z}(1)|\psi_{p_z}(1)\rangle$ will be written as overlap integrals, S_{11}:

$$\begin{aligned} N^2(S_{11}^2 + S_{22}^2 + S_{33}^2 + S_{44}^2 + S_{55}^2 + S_{66}^2) &= 1 \\ 6N^2 &= 1 \\ N &= \frac{1}{\sqrt{6}}. \end{aligned}$$

We can now write the wavefunction for the molecular orbital:

$$\psi_{\mathrm{MO}} = \frac{1}{\sqrt{6}}(\psi_{p_z}(1) + \psi_{p_z}(2) + \psi_{p_z}(3) + \psi_{p_z}(4) + \psi_{p_z}(5) + \psi_{p_z}(6)).$$

Exercise 71 Since the elements down any column (or along any row) add up to $x + 2$, all the columns are added to column 1, which makes all of its elements equal to $x + 2$. This can then be taken out as a factor, leaving

$$(x+2)\begin{vmatrix} 1 & 1 & 0 & 0 & 0 & 1 \\ 1 & x & 1 & 0 & 0 & 0 \\ 1 & 1 & x & 1 & 0 & 0 \\ 1 & 0 & 1 & x & 1 & 0 \\ 1 & 0 & 0 & 1 & x & 1 \\ 1 & 0 & 0 & 0 & 1 & x \end{vmatrix} = 0.$$

If column 1 is subtracted from columns 2 and 6 then there is only a single 1 in the first row. We can then use equation (5.20) to reduce the determinant to a 5×5 one:

$$(x+2)\begin{vmatrix} x-1 & 1 & 0 & 0 & -1 \\ 0 & x & 1 & 0 & -1 \\ -1 & 1 & x & 1 & -1 \\ -1 & 0 & 1 & x & 0 \\ -1 & 0 & 0 & 1 & x-1 \end{vmatrix} = 0.$$

If column 5 is added to column 2 then $(x-1)$ may be factorised out, leaving

$$(x+2)(x-1)\begin{vmatrix} x-1 & 0 & 0 & 0 & -1 \\ 0 & 1 & 1 & 0 & -1 \\ -1 & 0 & x & 1 & -1 \\ -1 & 0 & 1 & x & 0 \\ -1 & 1 & 0 & 1 & x-1 \end{vmatrix} = 0.$$

Next, columns 3 and 5 are added to column 1. This generates $(x - 2)$ as a common factor that can be taken out to give

$$(x+2)(x-2)(x-1)\begin{vmatrix} 1 & 0 & 0 & 0 & -1 \\ 0 & 1 & 1 & 0 & -1 \\ 1 & 0 & x & 1 & -1 \\ 0 & 0 & 1 & x & 0 \\ 1 & 1 & 0 & 1 & x-1 \end{vmatrix} = 0.$$

Next, column 1 is added to column 5 to leave only a single 1 in the first row. The determinant can then be reduced (equation (5.20)) to

$$(x+2)(x-2)(x-1)\begin{vmatrix} 1 & 1 & 0 & -1 \\ 0 & x & 1 & 0 \\ 0 & 1 & x & 0 \\ 1 & 0 & 1 & x \end{vmatrix} = 0.$$

Next, column 1 is subtracted from column 2 and added to column 4 to leave only a single 1 in the first row. This reduces the determinant to

$$(x+2)(x-2)(x-1)\begin{vmatrix} x & 1 & 0 \\ 1 & x & 0 \\ -1 & 1 & x+1 \end{vmatrix} = 0.$$

Column 2 is now added to column 1, giving $(x+1)$ as a common factor that can be taken out to give

$$(x+2)(x-2)(x+1)(x-1)\begin{vmatrix} 1 & 1 & 0 \\ 1 & x & 0 \\ 0 & 1 & x+1 \end{vmatrix} = 0.$$

Column 1 is then subtracted from column 2, leaving only a single 1 in the first row, reducing the determinant to

$$(x+2)(x-2)(x+1)(x-1)\begin{vmatrix} x-1 & 0 \\ 1 & x+1 \end{vmatrix} = 0.$$

The final determinant multiplies out to give $x^2 - 1$ which factorises to $(x+1)(x-1)$. Hence the original equation can be written as

$$(x+2)(x-2)(x+1)^2(x-1)^2 = 0,$$

which clearly has the solutions $x = \pm 1(\text{twice}), \pm 2$, as desired.

Chapter 6: Atomic Spectroscopy

Exercise 72 The parity-allowed transition are the ones between states of opposite parity: s–p, s–f, p–d and d–f. The parity-forbidden transitions are the same-parity ones: s–s, p–p, d–d, f–f, s–d, p–f. It is the violation of this selection rule that accounts for the fairly pale colour of d–d transitions in transition metal complex ions. More strongly coloured transition metal ions, such as manganate(VII) and chromate(VI), tend to gain their colour from a different type of electronic transition, such as ligand-to-metal-charge-transfer (LMCT) transitions.

Exercise 73 The ground state for Eu^{3+} f^6 is $^7F_0$ (see Table A.6). Hence this electric-dipole transition breaks all four selection rules.

Exercise 74 We need to fix N such that $\langle\psi(x)|\psi(x)\rangle = 1$. We don't need to worry about the complex conjugate as $\psi(x)$ is real.

$$\int_0^L N^2 \sin^2\left(\frac{n\pi x}{L}\right) \mathrm{d}x = 1$$

We solve the integral with the trigonometric identity $\sin^2 x = \frac{1}{2}(1 - \cos 2x)$:

$$\frac{N^2}{2}\int_0^L \left(1 - \cos\left(\frac{2n\pi x}{L}\right)\right) \mathrm{d}x = 1$$

$$\frac{N^2}{2}\left[x - \frac{L}{2n\pi}\sin\left(\frac{2n\pi x}{L}\right)\right]_0^L = 1$$

$$\frac{N^2}{2}.L = 1$$

$$N = \sqrt{\frac{2}{L}}.$$

Exercise 75 We need to show that $\langle\psi(n = 1)|\psi(n = 2)\rangle = 0$. We use the trigonometric identity $\sin 2x = 2\sin x\cos x$:

$$\begin{aligned}\int_0^L N^2 \sin\left(\frac{\pi x}{L}\right)\sin\left(\frac{2\pi x}{L}\right) \mathrm{d}x &= 2N^2\int_0^L \sin^2\left(\frac{\pi x}{L}\right)\cos\left(\frac{\pi x}{L}\right) \mathrm{d}x\\ &= 2N^2\left[\frac{L}{3\pi}\sin^3\left(\frac{\pi x}{L}\right)\right]_0^L\\ &= 0.\end{aligned}$$

Exercise 76 The significance of the trigonometric identity is that we can express the product of any two wavefunctions as two cosine function of a multiple of the angles. When we integrate the cosine functions we obtain sine functions of multiples of the angles which will integrate to zero, proving that all the wavefunctions are orthogonal:

$$\begin{aligned}&\int_0^L \sin\left(\frac{n_1\pi x}{L}\right)\sin\left(\frac{n_2\pi x}{L}\right) \mathrm{d}x\\ &= \int_0^L \frac{1}{2}\left\{\cos\left(\frac{(n_1 - n_2)\pi x}{L}\right) - \cos\left(\frac{(n_1 + n_2)\pi x}{L}\right)\right\} \mathrm{d}x\end{aligned}$$

$$= \frac{1}{2}\left[\frac{L}{(n_1 - n_2)\pi}\sin\left(\frac{(n_1 - n_2)\pi x}{L}\right) - \frac{L}{(n_1 + n_2)\pi}\sin\left(\frac{(n_1 + n_2)\pi x}{L}\right)\right]_0^L$$

$$= 0.$$

Exercise 77 We take the electric-dipole operator to be $\hat{x}$. We use the trigonometric identity $\sin 2x = 2\sin x\cos x$:

$$\begin{aligned}
\mu_{12} &= \langle\psi(n=1)\mid x\mid\psi(n=2)\rangle \\
&= \int_0^L x\sin\left(\frac{\pi x}{L}\right)\sin\left(\frac{2\pi x}{L}\right)\,\mathrm{d}x \\
&= \int_0^L 2x\sin^2\left(\frac{\pi x}{L}\right)\cos\left(\frac{\pi x}{L}\right)\,\mathrm{d}x.
\end{aligned}$$

We integrate by parts: $u = 2x$, $\mathrm{d}u = 2\mathrm{d}x$, $\mathrm{d}v = \sin^2\left(\frac{\pi x}{L}\right)\cos\left(\frac{\pi x}{L}\right)$, $v = \frac{L}{3\pi}\sin^3\left(\frac{\pi x}{L}\right)$.

$$\int 2x\sin^2\left(\frac{\pi x}{L}\right)\cos\left(\frac{\pi x}{L}\right)\,\mathrm{d}x = \frac{2L}{3\pi}\left\{x\sin^3\left(\frac{\pi x}{L}\right) - \int\sin^3\left(\frac{\pi x}{L}\right)\,\mathrm{d}x\right\}$$

To do the next integration we use the trigonometric identity $\sin^3 x = \frac{1}{4}(3\sin x - \sin 3x)$:

$$\begin{aligned}
\int_0^L \sin^3\left(\frac{\pi x}{L}\right)\,\mathrm{d}x &= \int_0^L\left\{\frac{3}{4}\sin\left(\frac{\pi x}{L}\right) - \frac{1}{4}\sin\left(\frac{3\pi x}{L}\right)\right\}\,\mathrm{d}x \\
&= \left[-\frac{3L}{4\pi}\cos\left(\frac{\pi x}{L}\right) + \frac{L}{12\pi}\cos\left(\frac{3\pi x}{L}\right)\right]_0^L.
\end{aligned}$$

We can now complete the integration:

$$\begin{aligned}
\mu_{12} &= \frac{2L}{3\pi}\left[x\sin^3\left(\frac{\pi x}{L}\right) + \frac{3L}{4\pi}\cos\left(\frac{\pi x}{L}\right) - \frac{L}{12\pi}\cos\left(\frac{3\pi x}{L}\right)\right]_0^L \\
&= \frac{2L}{3\pi}\left(-\frac{3L}{2\pi} + \frac{L}{6\pi}\right) \\
&= -\frac{8L^2}{9\pi^2}.
\end{aligned}$$

The finite transition moment indicates that the $n = 1 \rightarrow n = 2$ transition is allowed. Any excitation to an even n level involves a $\sin nx$ term representing the excited wavefunction, where n is even. These are then expressed as functions of $\sin x$ and $\cos x$, both raised to odd powers. After multiplying by the initial wavefunction the $\sin x$ term is raised to an even power. Integration of these functions leads to sine and cosine functions involving odd-number multiples of the angle. The cosine functions will always integrate to a finite result and so any parity-changing transition is allowed.

For the case of the $n = 1 \rightarrow n = 3$ transition, the $\sin 3x$ term is expressed in terms of $\sin^3 x$ as above. When multiplied by the other $\sin x$ term this gives a term in $\sin^4 x$. We then make use of the trigonometric identity $\sin^4 x = \frac{1}{8}(3 - 4\cos 2x + \cos 4x)$. We simplify the trigonometric functions as far as possible before the integration:

$$\begin{aligned}
x \sin x \sin 3x &= x(3\sin^2 x - 4\sin^4 x) \\
&= \frac{3x}{2}(1 - \cos 2x) - \frac{x}{2}(3 - 4\cos 2x + \cos 4x) \\
&= \frac{x}{2}(\cos 2x - \cos 4x).
\end{aligned}$$

Using integration by parts we can work out a useful general result:

$$\int_0^L \frac{x}{2}\cos\left(\frac{n\pi x}{L}\right) \mathrm{d}x = \left[\frac{x}{2}\left(\frac{L}{n\pi}\right)\sin\left(\frac{n\pi x}{L}\right) + \frac{1}{2}\left(\frac{L}{n\pi}\right)^2 \cos\left(\frac{n\pi x}{L}\right)\right]_0^L$$

The sine terms will lead to zero contributions for any integer value of n. The cosine terms, however, will only come to zero when n is even. This result shows both integrals that make up μ_{13} come to zero, confirming that the transition is forbidden. Any excitation from $n = 1$ to an odd value of n will produce a $\sin nx$ wavefunction for the excited state where n is odd. These can all be expressed in terms of simple functions involving odd powers of $\sin x$. On multiplication by the $\sin x$ of the ground state this gives a series of functions of $\sin x$ raised to even powers. All such functions can be expressed as cosine functions of multiple angles which integrate to functions of the sort above that come to zero, as we have seen. Hence any parity-conserving transition in the 1D box is forbidden.

Exercise 78 As there is no potential energy in the box, we can say the particle has a total energy of 1 J. We substitute values into equation (6.11) and obtain the particle's quantum number n:

$$n = \sqrt{\frac{8EmL^2}{h^2}} = \sqrt{\frac{8}{(6.626 \times 10^{-34})^2}} = 4.269 \times 10^{33}.$$

The energy gap to the next quantum level may be found using equation (6.12):

$$\begin{aligned}\Delta E &= \frac{h^2}{8mL^2}(2n+1) = \frac{(6.626 \times 10^{-34})^2}{8} \times 2 \times 4.269 \times 10^{33}\,\mathrm{J} \\ &= 4.685 \times 10^{-34}\,\mathrm{J}.\end{aligned}$$

This energy gap is too small to be observable; the energy states for the particle appear to be a continuum which, of course, is consistent with our everyday experience.

Exercise 79 We insert into equation (6.12) $L = Nl$ and $\Delta E = \frac{hc}{\lambda}$. Since the molecule will have $N+1$ carbon atoms, each inserting one electron into the π orbitals, we deduce that the electrons fill up to $n = \frac{N+1}{2}$.

$$N^2 l^2 = \frac{(N+2)h\lambda}{8mc}$$

The only unknown in this equation is N, so we can solve it quadratically. We can simplify the algebra if we divide through by l^2 and introduce $k = \frac{h\lambda}{8mcl^2} = 6.19$:

$$N^2 - kN - 2k = 0.$$

This equation has two roots:

$$N = \frac{k \pm \sqrt{k^2 + 4k}}{2} = 3.1 \pm 4.7.$$

We reject the negative root, leaving us with $N = 7.8$. A conjugated chain will consist of an odd number of C–C bonds, which we could take to be seven or nine, corresponding to eight or ten carbon atoms. In reality, the λ value first exceeds 400 nm at $N = 13$, but in practice the transitions span a broad wavelength range. Considering how crude the model is, it seems a surprisingly sensible answer. Our model takes no account of the fact there are even any atoms involved, let alone the properties of the atomic $2p_z$ orbitals that are involved in these delocalised systems.

Exercise 80 We apply equation (2.39):

$$\langle \psi \mid x \mid \psi \rangle = \int_0^{L_y} \int_0^{L_x} x \frac{2}{L_x} \sin^2 \left(\frac{n_x \pi x}{L_x} \right) \frac{2}{L_y} \sin^2 \left(\frac{n_y \pi y}{L_y} \right) dx dy.$$

The integral over y will simply be over the square of the y-part of the wavefunction, which will return 1 as the wavefunction is normalised. The important integral is the one over x, which will be over x times the square of the x-part of the wavefunction. We use the following integral (obtained though integration by parts):

$$\int x \sin^2 x \, dx = \frac{1}{4} x^2 - \frac{1}{4} x \sin 2x - \frac{1}{8} \cos 2x + c.$$

Integrating between 0 and L_x the trigonometric terms will vanish, leaving $\frac{1}{4} L_x^2$. This needs to be multiplied by the square of the normalisation constant, $\frac{2}{L}$, leaving $\frac{1}{2} L_x$ as the average of x, which we would expect as the wavefunction is symmetric around $\frac{1}{2} L_x$. Hence $(\langle x \rangle)^2 = \frac{1}{4} L_x^2$.

We take a similar approach to $\langle \psi \mid x^2 \mid \psi \rangle$. Again, we can ignore the y components. We use the following integral (which requires integration by parts twice):

$$\int x^2 \sin^2 x \, dx = \frac{1}{6} x^3 - \frac{1}{4} x^2 \sin 2x - \frac{1}{4} x \cos 2x + \frac{1}{8} \sin 2x + c.$$

Again, the trigonometric terms vanish when integrating between 0 and L_x. This leaves $\frac{1}{6} L_x^3$, which again is multiplied by the square of the normalisation constant to give $\frac{1}{3} L_x^2$ as the average of x^2. We see that this is different from $(\langle x \rangle)^2$, as desired.

Exercise 81 Since the operators $\hat{x}$ and $\hat{y}$ act independently of each other, and since the half of the wavefunction not being operated on integrates to 1 as it is normalised, we can use the results of the last question to show the following result:

$$\langle x \rangle \langle y \rangle = \langle xy \rangle = \frac{1}{4} L_x L_y.$$

Exercise 82 Since the ground state ($n_x = n_y = n_z = 1$) is at $3k$, we consider states up to $E = 15k$. The five excited energy levels are given in Table A.8. The parity is given by $(-1)^{n_x + n_y + n_z - 1}$, $+1$ denoting even and -1 denoting odd.

Table A.8. Excited states in a cubic box.

n_x	n_y	n_z	Energy/k	g	Parity
2	1	1	6	3	odd
2	2	1	9	3	even
2	2	2	12	1	odd
3	1	1	11	3	even
3	2	1	14	6	odd

Exercise 83 We saw in the last question that excitation to the first excited state involves $3k$ of energy. We equate this to $\frac{hc}{\lambda}$ and solve for L:

$$L^2 = \frac{3h\lambda}{8mc} = \frac{3 \times 6.63 \times 10^{-34}\,\mathrm{J\,s} \times 5.1 \times 10^{-7}\,\mathrm{m}}{8 \times 9.11 \times 10^{-31}\,\mathrm{kg} \times 3.00 \times 10^{8}\,\mathrm{m\,s^{-1}}}$$
$$= 4.64 \times 10^{-19}\,\mathrm{m}^2$$
$$L = 6.8 \times 10^{-10}\,\mathrm{m}.$$

Since bromide is larger than chloride, L will be larger so λ will be longer, i.e. redder. If redder light is being absorbed, the crystal will appear bluer.

Exercise 84 Equation (6.30) shows that the product of any combination of the possible sine and cosine functions gives the sum of two sine functions of multiples of π. They need to be integrated over all space which, for a ring, means integrating with respect to φ between the limits of 0 and 2π. The two sine functions will each integrate to cosine functions of the multiples of π. Integrating between 0 and 2π these will all come to zero, confirming that all the possible combinations of sine and cosine functions are orthogonal.

Exercise 85 The radius of the ring is $6 \times 1.40 \times 10^{-10}\,\mathrm{m}/2\pi = 1.34 \times 10^{-10}\,\mathrm{m}$. The three pairs of π electrons will fill the $m = 0$ and $m = \pm 1$ levels. So the lowest energy transition will be $m = \pm 1 \rightarrow m = \pm 2$, which is allowed by the parity and angular momentum selection rules.

$$\lambda = \frac{hc}{\Delta E_{1\rightarrow 2}} = \frac{8\pi^2 M_e r^2 c}{h(2m+1)}$$
$$= \frac{8\pi^2 \times 9.11 \times 10^{-31}\,\mathrm{kg} \times (1.34 \times 10^{-10}\,\mathrm{m})^2 \times 3.00 \times 10^{8}\,\mathrm{m\,s^{-1}}}{6.626 \times 10^{-34}\,\mathrm{J\,s} \times 3}$$
$$= 1.94 \times 10^{-7}\,\mathrm{m}$$

The predicted value of 194 nm is surprisingly close to the main observed peak at 184 nm considering that the particle-on-a-ring model includes no details of atoms or p_z orbitals. It is not surprising that additional features are seen in the experimental spectrum since this model neglects electron–electron interactions.

Exercise 86 First we apply the Schrödinger equation for a particle on a sphere to just the P_1^0 function, $\cos\theta$. We can therefore neglect the differential operator with respect to φ:

$$\begin{aligned}\frac{-\hbar^2}{2I}\left(\frac{1}{\sin\theta}\frac{\partial}{\partial\theta}\sin\theta\frac{\partial}{\partial\theta}\right)\cos\theta &= E\cos\theta\\ \frac{-\hbar^2}{2I}\frac{1}{\sin\theta}\frac{\partial}{\partial\theta}(-\sin^2\theta) &= E\cos\theta\\ \frac{-\hbar^2}{2I} - 2\cos\theta &= E\cos\theta\\ \frac{2\hbar^2}{2I} &= E.\end{aligned}$$

Next we apply the Schrödinger equation to the P_1^1 function, $\sin\theta$, together with the complex exponential $e^{i\varphi}$:

$$\begin{aligned}\frac{-\hbar^2}{2I}\left(\frac{1}{\sin\theta}\frac{\partial}{\partial\theta}\sin\theta\frac{\partial}{\partial\theta}+\frac{1}{\sin^2\theta}\frac{\partial^2}{\partial\varphi^2}\right)\sin\theta e^{i\varphi} &= E\sin\theta e^{i\varphi}\\ \frac{-\hbar^2}{2I}\frac{e^{i\varphi}}{\sin\theta}(\cos^2\theta-\sin^2\theta-1) &= E\sin\theta e^{i\varphi}\\ \frac{-\hbar^2}{2I}e^{i\varphi}. - 2\sin\theta &= E\sin\theta e^{i\varphi}\\ \frac{2\hbar^2}{2I} &= E.\end{aligned}$$

We see that the $|10\rangle$ and $|11\rangle$ eigenfunctions have the same energy. The only difference of $|1-1\rangle$ compared with $|11\rangle$ is the negative sign in the complex exponential. On application of the second derivative operator with respect to φ the minus sign disappears, and so the energy of the $|1-1\rangle$ eigenfunction is no different to the other eigenfunctions with $l = 1$.

Exercise 87 E is obviously zero for $|00\rangle$ since all the terms in the operator are differentials and there is no variable in the wavefunction.

We have already found that $E = \frac{2\hbar^2}{2I}$ for $|10\rangle$. For $|20\rangle$ we insert $P_2^0 = 3\cos^2\theta - 1$ into the Schrödinger equation, omitting the differential term with respect to φ:

$$\begin{aligned}
\frac{-\hbar^2}{2I}\left(\frac{1}{\sin\theta}\frac{\partial}{\partial\theta}\sin\theta\frac{\partial}{\partial\theta}\right)(3\cos^2\theta - 1) &= E(3\cos^2\theta - 1) \\
\frac{-\hbar^2}{2I}\frac{1}{\sin\theta}\frac{\partial}{\partial\theta}(-6\cos\theta\sin^2\theta) &= E(3\cos^2\theta - 1) \\
\frac{-\hbar^2}{2I}\frac{-6}{\sin\theta}(\cos\theta\cdot 2\sin\theta\cos\theta - \sin^2\theta\cdot\sin\theta) &= E(3\cos^2\theta - 1) \\
\frac{-\hbar^2}{2I}\cdot -6(2\cos^2\theta - \sin^2\theta) &= E(3\cos^2\theta - 1) \\
\frac{-\hbar^2}{2I}\cdot -6(3\cos^2\theta - 1) &= E(3\cos^2\theta - 1) \\
\frac{6\hbar^2}{2I} &= E.
\end{aligned}$$

We can see that for $l = 0, 1, 2$, their energies in units of $\frac{\hbar^2}{2I}$ of 0, 2 and 6, respectively, are what is predicted by the $l(l+1)$ expression.

Exercise 88 For the energy of transitions from l to $l+1$ we need to substitute $l+1$ into $l(l+1)$ and subtract $l(l+1)$:

$$\begin{aligned}
(l+1)(l+2) - l(l+1) &= l^2 + 3l + 2 - l^2 - l \\
&= 2l + 2 \\
&= 2(l+1).
\end{aligned}$$

For the difference between successive energies we substitute $l+1$ into the $2(l+1)$ and subtract $2(l+1)$:

$$\begin{aligned}
2(l+2) - 2(l+1) &= 2l + 4 - 2l - 2 \\
&= 2.
\end{aligned}$$

Exercise 89 For each value of l there are $2l+1$ eigenfunctions, each of which may have two electrons associated with it, making a total of $4l+2$ electrons taking each value of l, starting from the ground state, $l = 0$. The electrons therefore fill the energy levels as follows: 2(s), 6(p), 10(d), 14(f), 18(g), 22(h), ... Therefore 50 electrons represent a closed shell (fully occupied up to g), leaving the last 10 electrons among the 11 h orbitals with $l = 5$.

Exercise 90 If the 5-degenerate crystal-field level were lowest in energy then the ten h electrons would fill it, leaving no unpaired electrons, consistent with observation. In K_3C_{60}, the buckyball gains the valence electron from each potassium atom, which will half-fill the next 3-degenerate level, leaving these extra three electrons spin-parallel, and therefore the buckyball paramagnetic. In K_6C_{60} the buckyball gains six electrons from the potassium atoms, which fill the 3-degenerate level, leaving it all-spin-paired and diamagnetic again.

Exercise 91 Substituting the values of n and $Z = 1$ into equation (2.34) and taking the difference between the two energy levels gives $\Delta E = \frac{3}{4}R_H$. The factor Z^2 is 4 times larger for He compared to H, so the $1/n^2$ terms need to be 4 times smaller for the energy change to be the same. Since $2^2 = 4$, this is achieved when the values for n in He are twice as large as those for hydrogen. Hence the $n = 4 \rightarrow n = 2$ transition in helium has the same energy as the hydrogen Lyman α-line.

Exercise 92 If the electron that isn't excited shields the excited electron perfectly then $Z_{\text{eff}} = 1$. We rearrange the Rydberg equation:

$$n^* = Z_{\text{eff}}\sqrt{\frac{R_H}{E}}.$$

We plug the values into the Rydberg equation. For the ground state,

$$n^* = \sqrt{\frac{109679}{198305}} = 0.74.$$

For the first excited state in absorption,

$$n^* = \sqrt{\frac{109679}{27176}} = 2.01.$$

We see that the n^* value for the 1s2p excited singlet state is in very good agreement with the actual value of n, showing that the approximation $Z_{\text{eff}} = 1$ was valid. For the electron in the ground state, the n^* was significantly less than the true value of 1, reflecting the imperfect shielding by the other electron. This is consistent with the shielding electron being in an inner shell in the ^{1}P case and in the same shell in the ground state case.

Exercise 93 To use the Rydberg equation we measure the energy of the subshell from the ionisation limit. We rearrange the Rydberg equation (2.34) with $Z_{\text{eff}} = 1$ and make n^* the subject:

$$n^*_{3s} = \frac{109679}{41450} = 1.63$$

$$n^*_{3p} = \frac{109679}{41450 - 16968} = 2.12$$

$$n^*_{3d} = \frac{109679}{41450 - 29173} = 2.99$$

$$n^*_{4s} = \frac{109679}{41450 - 25740} = 2.64.$$

We now find δ_{nl} by equating it to $n - n^*$:

$$\delta_{3s} = 3 - 1.63 = 1.37$$

$$\delta_{3p} = 3 - 2.12 = 0.88$$

$$\delta_{3d} = 3 - 2.99 = 0.01$$

$$\delta_{4s} = 4 - 2.64 = 1.36.$$

The quantum defect is largest for s subshells, smaller for p and tiny for d. This is consistent with the penetration of these subshells and the number of radial nodes, $n - l - 1$. It is also noteworthy that 3s and 4s subshells have very similar quantum defects. Quantum defects generally vary little with n [94].

Exercise 94 The excited state will be $2s^2$ 2p 3d. The 2s electrons may be ignored as they constitute a closed subshell with no net angular momentum of any type. Considering the number of spin-orbitals in each subshell we can deduce that the 2p 3d configuration will be $6 \times 10 = 60$-fold degenerate. The highest orbital angular momentum state, $M_L = 3$ may have singlet or triplet spin multiplicity as the electrons are in different subshells. This gives ^{3}F and ^{1}F, which together account for 28 microstates. With no Pauli-principle restrictions, the terms may be deduced by the vector addition of the $l = 2$ and $l = 1$ orbital angular momentum vectors, and allowing the resultant to be both a triplet and a singlet term. Hence the other terms will be ^{3}D and ^{1}D, and ^{3}P and ^{1}P, accounting for 15, 5, 9 and 3 microstates, respectively. All the microstates sum to 60, confirming that the conclusion is correct. Hund's rules predict that ^{3}F will be the lowest energy term. It is observed to be the second lowest in energy (after ^{1}D). Hund's rules are less reliable with excited configurations as often, as in this case, there are other terms close by energetically from other configurations, and second-order interactions between them can modify the anticipated ordering of terms.

Exercise 95 Since the two electrons in this microstate have antiparallel spins, the Kronecker delta in the exchange integral makes it vanish. Hence the electrostatic energy of interaction between the two electrons is equal to the direct-coulombic integral, J_{12}. The $c^k(+1, +1)$ integrals for p electrons come to 1 for $k = 0$ and $-\frac{1}{5}$ for $k = 2$ (shown in Table 6.2). Since both electrons of the microstate have the same m_l value, we need to square the c^k results. Terms involving different ranks are independent, so we don't consider cross terms. The product of the c^k integrals comes to 1 for $k = 0$ and $+\frac{1}{25}$ for $k = 2$. We can now find the direct-coulombic interaction energy, J_{12}, and therefore the total electrostatic interaction energy:

$$J_{12} = F^0 + \frac{1}{25}F^2 = F_0 + F_2.$$

All the microstates within a term are degenerate so $F_0 + F_2$ is the energy of the ^{1}D term.

Exercise 96 The $M_S = 1, M_L = 1$ state of ^{3}P is composed of just a single microstate, 1^+0^+ so the energy of ^{3}P will be equal to the energy of this microstate. The angular part of the direct-coulombic interaction energy is $c^k(+1, +1)c^k(0, 0)$. Results from Table 6.2 give the direct-coulombic energy as $F_0 - 2F_2$. The angular part of the exchange-coulombic interaction energy is $(c^k(+1, 0))^2$, giving the overall exchange-coulombic energy as $3F_2$. The total electrostatic energy of interaction is $J - K$, which comes to $F_0 - 5F_2$.

Exercise 97 We equate the sum of the energies of the three microstates with $M_S = M_L = 0$ to the sum of the energies of the three terms of p^2. All of the microstates have antiparallel electrons so there are no exchange-coulombic energies to consider. We use the results in Table 6.2 to find the direct-coulombic energies of the $(1^+, -1^-)$, $(1^-, -1^+)$ and $(0^+, 0^-)$ microstates. The first two microstates have the same energy, $F_0 + F_2$, since the electron spin doesn't affect the direct-coulombic integrals. The final microstate has the direct-coulombic energy $F_0 + 4F_2$. The sum of the energies of the three microstates therefore comes to $3F_0 + 6F_2$. This is equated to the sum of the p^2 term energies:

$$\begin{aligned} 3F_0 + 6F_2 &= E(^1\mathrm{D}) + E(^3\mathrm{P}) + E(^1\mathrm{S}) \\ &= (F_0 + F_2) + (F_0 - 5F_2) + E(^1\mathrm{S}) \\ E(^1\mathrm{S}) &= F_0 + 10F_2. \end{aligned}$$

The Slater F_k parameters are always positive. Therefore, these results put ^{3}P lowest in energy and ^{1}S highest in energy, which is consistent with the principles of Hund's rules.

The energy gaps between the different p^2 terms in this model depend only on the F_2 coefficient as all the terms share a common F_0 term in their energy. The energy gap between ^{1}S and ^{1}D is $9F_2$, while the gap between ^{1}D and ^{3}P is $6F_2$, confirming the relative spacings given in the question. This ratio, $[E(^1\mathrm{S})-E(^1\mathrm{D})]$:$[E(^1\mathrm{D})-E(^3\mathrm{P})]$, of 1.5 does not predict the actual spacings accurately. The observed ratios for p^2 in C, N^+, O^{2+}, ..., Mg^{6+} are all between 1.12 and 1.14 [63]. This is probably due to interaction with the same-parity 2p 3p configuration [29].

Exercise 98 Fortunately the energy of the ground and first excited terms can each both be calculated from a single three-electron microstate. The microstates of the p^3 configuration are collected in Table A.2. The microstate $(1^+, 0^+, -1^+)$ gives the energy of ^{4}S, while $(1^+, 1^-, 0^+)$ gives the energy of ^{2}D. We need to consider the three-electron microstates as a sum of three two-electron interactions:

$$\begin{aligned} (1^+, 0^+, -1^+) &= (1^+, 0^+) + (1^+, -1^+) + (0^+, -1^+) \\ (1^+, 1^-, 0^+) &= (1^+, 1^-) + (1^+, 0^+) + (1^-, 0^+). \end{aligned}$$

Each of the above two-electron interaction energies are found using the methods described for p^2. All three two-electron interactions for the ^{4}S calculation involve exchange energies as all the electrons are parallel, whereas only the second of three two-electron interactions for ^{2}P involves an exchange interaction. The following results are obtained:

$$\begin{aligned} E(^4\mathrm{S}) &= 3F_0 - 15F_2 \\ E(^2\mathrm{D}) &= 3F_0 - 6F_2. \end{aligned}$$

Therefore the difference in energy between the two terms is $9F_2$. Note that both term energies involve $3F_0$. This reflects the three pairs of electron that can be chosen in p^3.

Exercise 99 The earlier exercises have established that the energy gap between ground and first-excited terms in p^2 and p^3 to be $6F_2$ and $9F_2$, respectively. Owing to the similarity between p^n and p^{6-n} configurations, there is also a $6F_2$ energy gap between ground and first-excited terms in p^4. Hence the F_2 values, in wavenumbers, for each element are: 1694 (C), 2136 (N) and 2632 (O). These values suggest an increase in the F_2 parameter and an increase in electrostatic repulsion energy across the period, consistent with the decreasing atomic radius and increasing number of valence electrons.

No analogous F_2 value can be assigned to boron as it only has a single p electron, so there can be no electron–electron interactions in its valence 2p subshell. Fluorine, p^5, by contrast, has multiple p electrons but, being spectroscopically equivalent to p^1, has only a single term, ^{2}P, and so scope for using an energy difference between terms to cancel out an F_0 parameter.

Exercise 100 We substitute $J-1$ into equation (6.49) and subtract it from the equation in terms of J:

$$\Delta E_{\text{s-o}} = \lambda_{SL}\frac{1}{2}[J(J+1) - J(J-1)] = \lambda_{SL}J.$$

Exercise 101 Iron has the ground configuration $3d^6\ 4s^2$. We ignore the closed 4s subshell and maximise the spin multiplicity of d^6 (Hund's first rule). The configuration has just one quintet term, which is the ground state, ^{5}D. It will therefore be described by the $(2^+, 2^-, 1^+, 0^+, -1^+, -2^+)$ microstate, which is the only microstate with $M_S = 2$ and $M_L = 2$. The electron–hole symmetry with the d^4 configuration means that we can proceed with d^4 (which will shorten the calculation) using the same terms, but will have to change the signs of ζ_{3d} and $\lambda_{^5\text{D}}$. The analogous d^4 microstate is $(2^+, 1^+, 0^+, -1^+)$. The sign change of the parameters will reverse the ordering of the J levels in energy so that the ground-state level is the one with maximum J. First we sum the spin–orbit energies of the electrons in the d^4 microstate using equation (6.47):

$$E_{\text{s-o}} = \zeta_{3d}\left(2.\frac{1}{2} + 1.\frac{1}{2} + 0.\frac{1}{2} - 1.\frac{1}{2}\right) = \zeta_{3d}.$$

We then work out the spin–orbit energy of $^5\text{D}_4$ in terms of $\lambda_{^5\text{D}}$ using equation (6.49). We use $J = 4$ since $M_J = M_S + M_L = 4$ for the microstate:

$$E_{\text{s-o}} = \lambda_{^5\text{D}}\frac{1}{2}(4\times 5 - 2\times 3 - 2\times 3) = 4\lambda_{^5\text{D}}.$$

Since the two calculated spin–orbit energies above are equal, we deduce that $\lambda_{^5D} = \frac{1}{4}\zeta_{3d}$.

Exercise 102 We found in the last exercise that $E_{\text{s-o}}(^5\text{D}_4) = 4\lambda_{^5\text{D}}$. Using the Landé interval rule, we deduce the spin–orbit energies of the remaining J levels:

$$E_{\text{s-o}}(^5\text{D}_3) = 0$$
$$E_{\text{s-o}}(^5\text{D}_2) = -3\lambda_{^5\text{D}}$$
$$E_{\text{s-o}}(^5\text{D}_1) = -5\lambda_{^5\text{D}}$$
$$E_{\text{s-o}}(^5\text{D}_0) = -6\lambda_{^5\text{D}}.$$

It appears that $^5\text{D}_0$ is the level with the lowest energy, which is the case for d^4 but, due to the sign change of the spin–orbit parameter from the electron–hole symmetry, $^5\text{D}_4$ is the ground level in d^6. The average spin–orbit energy, $E_{\text{s-o,av}}$, is found by weighting all the energies with the $(2J+1)$-fold degeneracy of each level, summing them and dividing by the total degeneracy, $(2S+1)(2L+1) = 25$:

$$E_{\text{s-o,av}} = \lambda_{^5\text{D}}\frac{1}{25}(9 \times 4 + 7 \times 0 - 5 \times 3 - 3 \times 5 - 1 \times 6) = 0.$$

This result confirms that the spin–orbit interaction splits the term into different levels without a shift in barycentre energy.

Exercise 103 Using the Landé interval rule, we subtract the observed energy of the $(J-1)$ level from the J level and set the result to $\lambda_{^5D}J$. We can use our earlier result that $\lambda_{^5\text{D}} = \frac{1}{4}\zeta_{3d}$ to set the interval to $\frac{1}{4}\zeta_{3d}J$. The intervals are all negative, which is consistent with the sign change of the spin–orbit parameter in more-than-half-full subshells. The four intervals produce the following results:

$$\zeta_{3d} = E_{\text{s-o}}(^5\text{D}_4) - E_{\text{s-o}}(^5\text{D}_3) \times 4/4 = -416\ \text{cm}^{-1}$$
$$\zeta_{3d} = E_{\text{s-o}}(^5\text{D}_3) - E_{\text{s-o}}(^5\text{D}_2) \times 4/3 = -384\ \text{cm}^{-1}$$
$$\zeta_{3d} = E_{\text{s-o}}(^5\text{D}_2) - E_{\text{s-o}}(^5\text{D}_1) \times 4/2 = -368\ \text{cm}^{-1}$$
$$\zeta_{3d} = E_{\text{s-o}}(^5\text{D}_1) - E_{\text{s-o}}(^5\text{D}_0) \times 4/1 = -360\ \text{cm}^{-1}.$$

While the values obtained for ζ_{3d} are of the same order, they are not the same, which is what the Landé interval rule and our theory of spin–orbit coupling require.

Exercise 104 The spin–orbit energy for ^{3}F can be determined directly from the microstate $(2^+, 1^+)$. We apply equations (6.47) and (6.49):

$$E_{\text{s-o}} = \zeta_{3d}(2.\frac{1}{2} + 1.\frac{1}{2}) = \frac{3}{2}\zeta_{3d}$$

$$E_{\text{s-o}} = \lambda_{^3\text{F}}\frac{1}{2}(4 \times 5 - 1 \times 2 - 3 \times 4) = 3\lambda_{^3\text{F}}$$

$$3\lambda_{^3\text{F}} = \frac{3}{2}\zeta_{3d}$$

$$\lambda_{^3\text{F}} = \frac{1}{2}\zeta_{3d}.$$

Two microstates correspond to $M_S = M_L = 1$: $(1^+, 0^+)$ and $(2^+, -1^+)$. We use the diagonal sum rule by applying equations (6.47) and (6.49) to the sum of the microstates:

$$E_{\text{s-o}} = \zeta_{3d}(1 \cdot \frac{1}{2} + 0 \cdot \frac{1}{2} + 2 \cdot \frac{1}{2} - 1 \cdot \frac{1}{2}) = \zeta_{3d}$$

$$E_{\text{s-o}} = (\lambda_{^3\text{F}} + \lambda_{^3\text{P}})\frac{1}{2}(2 \times 3 - 1 \times 2 - 1 \times 2) = \lambda_{^3\text{F}} + \lambda_{^3\text{P}}$$

$$\lambda_{^3\text{F}} + \lambda_{^3\text{P}} = \zeta_{3d}$$

$$\lambda_{^3\text{P}} = \frac{1}{2}\zeta_{3d}.$$

We see that the spin–orbit parameter for both multiplets is $\frac{1}{2}\zeta_{3d}$.

Exercise 105 We found in the previous exercise that the spin–orbit parameters for both multiplets were $\frac{1}{2}\zeta_{3d}$. We apply the Landé interval rule to both intervals in both multiplets:

$$\zeta_{3d} = E_{\text{s-o}}(^3\text{F}_4) - E_{\text{s-o}}(^3\text{F}_3) \times 2/4 = +108 \text{ cm}^{-1}$$

$$\zeta_{3d} = E_{\text{s-o}}(^3\text{F}_3) - E_{\text{s-o}}(^3\text{F}_2) \times 2/3 = +113 \text{ cm}^{-1}$$

$$\zeta_{3d} = E_{\text{s-o}}(^3\text{P}_2) - E_{\text{s-o}}(^3\text{P}_1) \times 2/2 = +110 \text{ cm}^{-1}$$

$$\zeta_{3d} = E_{\text{s-o}}(^3\text{P}_1) - E_{\text{s-o}}(^3\text{P}_0) \times 2/1 = +112 \text{ cm}^{-1}.$$

Here we see better agreement within and between the multiplets than was observed with ^{5}D in iron. The Z^4 relationship predicts that the ζ_{3d} parameter in iron should be larger than in titanium by a factor of $26^4/22^4 = 1.95$. The observed factor was more like 3.5.

Exercise 106 The p^3 configuration has the terms ^{4}S, ^{2}D and ^{2}P, as was found in Section 4.3. All S terms are unaffected by spin–orbit coupling, as is clear from equation (6.49). The microstate $(1^+, 1^-, 0^+)$ is the only one associated with the quantum numbers $M_S = \frac{1}{2}$ and $M_L = 2$ and so can be assigned to the term ^{2}D. We apply equation (6.47):

$$E_{\text{s-o}} = \zeta_p \left(1.\frac{1}{2} - 1.\frac{1}{2} + 0.\frac{1}{2} \right) = 0.$$

Therefore the ^{2}D term is not split by spin–orbit coupling. There are two microstates associated with the quantum numbers $M_S = \frac{1}{2}$ and $M_L = 1$: $(1^+, 0^+, 0^-)$ and $(1^+, 1^-, -1^+)$. Again we apply equation (6.47) to the sum of both microstates:

$$E_{\text{s-o}} = \zeta_p \left(1 \cdot \frac{1}{2} + 1 \cdot \frac{1}{2} + 1 \cdot \frac{-1}{2} - 1 \cdot \frac{1}{2} \right) = 0.$$

Therefore the ^{2}P term is also not split by spin–orbit coupling. Indeed the symmetry of the microstates of a half-full subshell is such that there can be no spin–orbit splitting of any of the terms. It is therefore curious that the ^{2}D term of nitrogen ($2p^3$) is observed to be split by 8 cm^{-1} [31], reflecting a breakdown of the model.

Bibliography

1. Janert, P.K., *Gnuplot in Action*, Manning, Greenwich–U.S.A., 2010.
2. Moore, B.G.: *Orbital plots using gnuplot*, Journal of Chemical Education, **77** (6), (2000), 785–9.
3. Griffiths, D.F. and Higham, D.J., *Learning LaTeX*, Society for Industrial and Applied Mathematics, 1997.
4. Grätzer, G., *More Math Into LaTeX*, Springer, New York–U.S.A. 4th edition, 2007.
5. Marinov, A. et al.: *Evidence for a long-lived superheavy nucleus with atomic mass number $A=292$ and atomic number $Z \cong 122$ in natural Th*, Int. J. Mod. Phys. E, **19** (1), (2010), 131–40.
6. Pyykkö, P.: *A suggested periodic table up to $Z \leq 172$, based on Dirac–Fock calculations on atoms and ions*, Phys. Chem. Chem. Phys., **13**, (2011), 161–8.
7. Dirac, P.A.M., *The Principles of Quantum Mechanics*, Oxford University Press, Oxford–U.K., 4th edition, 1958.
8. Cullerne, J.P. and Machacek, A.C., *The Language of Physics*, Oxford University Press, Oxford–U.K., 2008.
9. Pilar, F.L., *Elementary Quantum Chemistry*, McGraw–Hill, New York–U.S.A., 2nd edition, 1990.
10. Atkins, P.W. and Friedman, R.S., *Molecular Quantum Mechanics*, Oxford University Press, Oxford–U.K., 3rd edition, 1997.
11. Pusey, M.F., Barrett, J. and Rudolph, T.: *On the reality of the quantum state*, Nature Physics, **8**, (2012), 476–9.
12. Mulliken, R.S.: *Electronic structures of polyatomic molecules and valence. II. General considerations*, Physical Review, **41** (1), (1932), 49–71.
13. Mulder, P.: *Are orbitals observable?*, HYLE Int. J. Phil. Chem, **17** (1), (2011), 24–35.
14. Smeenk, C.T.L.: *Viewpoint: A new look at the hydrogen wave function*, Physics, **6**, (2013), 58.

15. Stodolna, A.S. et al.: *Hydrogen atoms under magnification: direct observation of the nodal structure of stark states*, Physical Review Letters, **110**, (2013), 213001.
16. Zuo, J.M., Kim, M., O'Keeffe, M. and Spence, J.C.H.: *Direct observation of d-orbital holes and Cu–Cu bonding in Cu_2O*, Nature, **40**, (1999), 49–52.
17. Scerri, E.R.: *Have orbitals really been observed?*, Journal of Chemical Education, **77** (11), (2000), 1492–4.
18. Wang, S.G. and Schwarz, W.H.E.: *On closed-shell structures, polar covalences, d shell holes and direct images of orbitals: The case of cuprite*, Angewandte Chemie Int. Ed., **39** (10), (2000), 1757–62.
19. Wang, S.G. and Schwarz, W.H.E.: *Final comment on the discussions of "The case of cuprite"*, Angewandte Chemie Int. Ed., **39** (21), (2000), 3794–6.
20. Itatani, J. et al.: *Tomographic imaging of molecular orbitals*, Nature, **432** (7019), (2004), 867–71.
21. Gross, L.: *Recent advances in submolecular resolution with scanning probe microscopy*, Nature Chemistry, **3** (4), (2011), 273–8.
22. Lüftner, D. et al.: *Imaging the wavefunctions of adsorbed molecules*, P. N. A. S., **111** (2), (2014), 605–10.
23. Schwarz, W.H.E.: *Measuring orbitals: Provocation or reality?*, Angewandte Chemie Int. Ed., **45** (10), (2006), 1508–17.
24. Dauth, M.: *Orbital density reconstruction for molecules*, Physical Review Letters, **107** (19), (2011), 193002.
25. Pauling, L. and Wilson, E.B., *Introduction to Quantum Mechanics*, McGraw–Hill, New York–U.S.A., 1935.
26. Powell, R.E.: *The five equivalent d orbitals*, Journal of Chemical Education, **45** (1), (1968), 45–8.
27. Ballhausen, C.J., *Introduction to Ligand Field Theory*, McGraw–Hill, New York–U.S.A., 1962.
28. Griffith, J.S., *The Theory of Transition-Metal Ions*, Cambridge University Press, Cambridge–U.K., 1961.
29. Condon, E.U. and Shortley, G.H., *The Theory of Atomic Spectra*, Cambridge University Press, Cambridge–U.K., 1963.
30. de Broglie, L., *The Revolution in Physics*, Routledge & Kegan Paul, London–U.K., 1954.
31. Moore, C.E., *Atomic Energy Levels As Derived from the Analyses of Optical Spectra, Volume 1, Hydrogen to Vanadium*, Circular of the National Bureau of Standards 467, Washington, D.C.–U.S.A., 1949.
32. Ball, P.: *Would element 137 really spell the end of the Periodic Table?*, Chemistry World, **7** (11), (2010), 35.
33. Bent, H.B. and Weinhold, F.: *News from the Periodic Table: An introduction to 'periodicity symbols, tables, and models for higher-order valency and donor–acceptor kinships'*, Journal of Chemical Education, **84** (7), (2007), 1145–6.
34. Vanquickenborne, L.G., Pierloot, K. and Devoghel, D.: *Transition metals and the aufbau principle*, Journal of Chemical Education, **71** (6), (1994), 469–71.

35. Scerri, E.R.: *Transition metal configurations and limitations of the orbital approximation*, Journal of Chemical Education, **66** (6), (1989), 481–3.
36. Melrose, M.P. and Scerri, E.R.: *Why the 4s orbital is occupied before the 3d*, Journal of Chemical Education, **73** (6), (1996), 498–503.
37. Slater, J.C.: *Atomic shielding constants*, Physical Review, **36** (1), (1930), 57–64.
38. Wulfsberg, G., *Inorganic Chemistry*, University Science Books, Sausalito–U.S.A., 2000, p. 19–20.
39. Ellis, H. (ed.), *Book of data (Revised Nuffield Advanced Science)*, Longman, Harlow–U.K., Revised edition, 1984, p. 44–5.
40. Winter, M., http://www.webelements.com. Accessed 29 March 2014.
41. Moore, C.E., *Atomic Energy Levels As Derived from the Analyses of Optical Spectra, Volume 2, Chromium to Niobium*, Circular of the National Bureau of Standards 467, Washington, D.C.–U.S.A., 1952.
42. Moore, C.E., *Atomic Energy Levels As Derived from the Analyses of Optical Spectra, Volume 3, Molybdenum to Lanthanum and Hafnium to Actinium*, Circular of the National Bureau of Standards 467, Washington, D.C. U.S.A., 1958.
43. Chen, E.C.M. and Wentworth, W.E.: *The experimental values of atomic electron affinities: Their selection and periodic behaviour*, Journal of Chemical Education, **52** (8), (1975), 486–9.
44. Andersen, T., Haugen, H.K. and Hotop, H.: *Binding energies in atomic negative ions*, Journal of Physical and Chemical Reference Data, **28** (6), (1999), 1511–33.
45. Wheeler, J.C.: *Electron affinities of the alkaline earth metals and the sign convention for electron affinity*, Journal of Chemical Education, **74** (1), (1997), 123–7.
46. Myers, R.T.: *The periodicity of electron affinity*, Journal of Chemical Education, **67** (4), (1990), 307–8.
47. Cohen–Tannoudji, C., Diu, B. and Laloë, F., *Quantum Mechanics, Volume 1*, Wiley–Interscience, Singapore, 2005.
48. Rowlands, P., *Zero to Infinity: The Foundations of Physics*, World Scientific, Singapore, 2007, p. 29.
49. Cohen–Tannoudji, C., Diu, B. and Laloë, F., *Quantum Mechanics, Volume 2*, Wiley–Interscience, Singapore, 2005.
50. Mackay, K.M., Mackay, R.A. and Henderson, W., *Introduction to Modern Inorganic Chemistry*, Stanley Thornes, Cheltenham–U.K., 5th edition, 1996, p. 132.
51. Mingos, D.M.P., *Essential Trends in Inorganic Chemistry*, Oxford University Press, Oxford–U.K., 1998, p. 276.
52. DeVault, D.: *A method of teaching the electronic structure of the atom II. Advanced topics*, Journal of Chemical Education, **21** (12), (1944), 575–81.
53. Keller, R.N.: *Energy level diagrams and extranuclear building of the elements*, Journal of Chemical Education, **39** (6), (1962), 289–93.
54. Pilar, F.L.: *4s is always above 3d!*, Journal of Chemical Education, **55** (1), (1978), 2–6.

55. Rich, R.L. and Suter, R.W.: *Periodicity and some graphical insights on the tendency toward empty, half-full and full subshells*, Journal of Chemical Education, **65** (8), (1988), 702–4.
56. Orofino, H., Machado, S.P. and Faria, R.B.: *The use of Rich and Suter diagrams to explain the electron configurations of transition elements*, Química Nova, **36** (6), (2013), 894–6.
57. Blake, A.B.: *Exchange stabilisation and the variation of ionization energy in the p^n and d^n series*, Journal of Chemical Education, **58** (5), (2013), 393–8.
58. Johnson, D.A., *Some thermodynamic aspects of inorganic chemistry*, Cambridge University Press, Cambridge–U.K., 2nd edition, 1982, p. 152–6.
59. Wang, S.G. and Schwarz, W.H.E.: *Icon of chemistry: the periodic system of chemical elements in the new century*, Angewandte Chemie Int. Ed., **48**, (2009), 2–14.
60. Kramida, A., Ralchenko, Y., Reader, J. and NIST ASD Team (2012). NIST Atomic Spectra Database (version 5.0), http://physics.nist.gov/asd. National Institute of Standards and Technology, Gaithersburg, MD–U.S.A.
61. Wang, S.G., Qiu, Y.X., Fang, H. and Schwarz, W.H.E.: *The challenge of the so-called electron configurations of the transition metals*, Chem. Eur. J., **12**, (2006), 4101–14.
62. Jensen, W.B.: *The positions of lanthanum (actinium) and lutetium (lawrencium) in the periodic table*, Journal of Chemical Education, **59** (8), (1982), 634–6.
63. Kuhn, H.G., *Atomic Spectra*, Longmans, London–U.K., 1962, p. 317–9.
64. Wilkins, R.G., *Kinetics and Mechanism of Reactions of Transition Metal Complexes*, VCH, Weinheim–Germany, 2nd edition, 1991, p. 199.
65. Calvo, F. et al.: *Evidence for low-temperature melting of mercury owing to relativity*, Angewandte Chemie Int. Ed., **52**, (2013), 7583–5.
66. Eichler, R. et al.: *Chemical characterization of element 112*, Nature, **447** (7140),(2007), 72–5.
67. Hoffman, D.C., Lee, D.M. and Pershina, V., *Transactinide Elements and Future Elements* in Morss, L. R., Edelstein, N. M. and Fuger, J, *The Chemistry of the Actinide and Transactinide Elements, volume 3*, 3rd edition, Dordrecht–The Netherlands, 2006, p. 1652–1752.
68. Pyykkö, P.: *Theoretical chemistry of gold*, Angewandte Chemie Int. Ed., **43** (116), (2004), 4412–56.
69. Martin, W.C., Zalubas, R. and Hagan, L., *Atomic Energy Levels — The Rare Earth Elements*, NSRDS-NBS 60, Washington, D.C.–U.S.A., 1978.
70. Drake, G.W.F. (ed.), *Atomic, Molecular and Optical Physics Handbook*, American Institute of Physics, Woodbury, New York–U.S.A., 1996.
71. Shannon, R.D.: *Revised effective ionic radii and systematic studies of interatomic distances in halides and chalcogenides*, Acta Crystallographica A, **32**, (1976), 751–67.
72. Wybourne, B.G., *Spectroscopic Properties of Rare Earths*, John Wiley & Sons, New York–U.S.A., 1965, p. 15.

73. Cowan, R.D., *The Theory of Atomic Spectra and Structure*, University of California Press, Berkeley–U.S.A., 1981.
74. Shaik, S. et al.: *Quadruple bonding in C_2 and analogous eight-valence electron species*, Nature Chemistry, **4** (3), (2012), 195–200.
75. Grunenberg, J.: *Quadruply bonded carbon*, Nature Chemistry, **4** (3), (2012), 154–5.
76. Schmidt, M.W., Ivanic, J. and Ruedenberg, K.: *Covalent bonds are created by the drive of electron waves to lower their kinetic energy through expansion*, J. Chem. Phys., **140**, (2014), 204104-1–14.
77. Cotton, F.A. et al.: *Mononuclear and polynuclear chemistry of rhenium(III): Its pronounced homophilicity*, Science, **145** (3638), (1964), 1305–7.
78. Nguyen, N. et al.: *Synthesis of a stable compound with fivefold bonding between two chromium(I) centers*, Science, **310** (5749), (2005), 844–7.
79. Minasian, S.G. et al.: *New evidence for 5f covalency in actinocenes determined from carbon K-edge XAS and electronic structure theory*, Chem. Sci., **5**, (2014), 351–9.
80. Potts, A.W. and Price, W.C.: *The photoelectron spectra of methane, silane, germane and stannane*, Proc. Roy. Soc. London A, **326** (1565), (1972), 176–79.
81. Greenwood, N.N. and Earnshaw, A., *Chemistry of the Elements*, Butterworth–Heinemann, Oxford–U.K., 2nd edition, 1997.
82. Coulson, C.A.: *The electronic structure of some polyenes and aromatic molecules. IV. The nature of the links of certain free radicals*, Proc. Roy. Soc. London A, **164** (918), (1938), 383–96.
83. Albright, T.A., Burdett, J.K. and Whangbo, M.-H., *Orbital Interactions in Chemistry*, John Wiley & Sons, New York–U.S.A., 1985.
84. Yoon, Z.S., Osuka, A. and Kim, D.: *Möbius aromaticity and antiaromaticity in expanded porphyrins*, Nature Chemistry, **1** (5), (2009), 113–22.
85. Huang, W. et al.: *A concentric planar doubly π-aromatic B_{19}^- cluster*, Nature Chemistry, **2** (3), (2010), 202–6.
86. Graybeal, J.D., *Molecular Spectroscopy*, McGraw–Hill Book Company, New York–U.S.A., 1988.
87. Nadje–Perge, S. et al.: *Spin–orbit qubit in a semiconductor nanowire*, Nature, **468** (7327), (2010), 1084-87.
88. Williams, D.H. and Fleming, I., *Spectroscopic Methods in Organic Chemistry*, McGraw–Hill Book Company, London–U.K., 4th edition, 1989, p. 18.
89. Crommie, M.F., Lutz, C.P. and Eigler, D.M.: *Confinement of electrons to quantum corrals on a metal surface*, Science, **262** (5131), (1993), 218–20.
90. Robinett, R.W.: *Visualising the solutions for the circular infinite well in quantum and classical mechanics*, American Journal of Physics, **64** (4), (1996), 440–6.
91. Harrison, P.M.C. and McCaw, C.S.: *Symmetry of buckminsterfullerene*, Education in Chemistry, **48** (4), (2011), 112–5.

92. Bohr, N.: *On the constitution of atoms and molecules. Part I*, Philosophical Magazine, **26** (151), (1913), 1–24.
93. Sannigrahi, A.B. and Das, R.: *Simple derivation of some basic selection rules*, Journal of Chemical Education, **57** (11), (1980), 786–8.
94. Softley, T.P., *Atomic Spectra*, Oxford University Press, Oxford–U.K., 1994, p. 45.
95. Rotenberg, M. et al., *The 3-j and 6-j symbols*, The Technology Press, M. I. T., MA–U.S.A., 1959.
96. Judd, B.R., *Operator Techniques in Atomic Spectroscopy*, Princeton University Press, NJ–U.S.A., 1998.
97. McCaw, C.S., Murdoch, K.M. and Denning, R.G.: *Energy levels of terbium(III) in the elpasolite* $Cs_2NaTbBr_6$*. I. Luminescence and two-photon excitation spectroscopy*, Molecular Physics, **101** (3), (2003), 427–38.
98. McCaw, C.S. and Denning, R.G.: *Energy levels of terbium(III) in the elpasolite* $Cs_2NaTbBr_6$*. II. A correlation crystal field analysis*, Molecular Physics, **101** (3), (2003), 439–47.

Index

Printed in the United States
By Bookmasters